BrightRED Study Guide

 **ADVANCED** Higher

BIOLOGY

David Lloyd and Geoff Morgan

First published in 2016 by:
Bright Red Publishing Ltd
1 Torphichen Street
Edinburgh
EH3 8HX

A CIP record for this book is available from the British Library

ISBN 978-1-906736-70-5

With thanks to:
Dr Neil Marshall, for his extensive contribution to the Investigative Biology chapter,
Leah McDowell, ELEM design, and Owen Rixon, The 2D Workshop (Illustrations),
PDQ Digital Media Solutions Ltd (layout) and Dr Anna Clark (edit)

Cover design and series book design by Caleb Rutherford – e i d e t i c

Acknowledgements
Every effort has been made to seek all copyright holders. If any have been overlooked, then Bright Red Publishing will be delighted to make the necessary arrangements.

Permission has been sought from all relevant copyright holders and Bright Red Publishing are grateful for the use of the following:

Acknowledgements
Permission has been sought from all relevant copyright holders and Bright Red Publishing are grateful for the use of the following:
© Mystrica Ltd. (p 9); Friedrich Fröbel (CC BY-SA 3.0)[1] (p 10); piemmea (CC BY-SA 3.0)[1] (p 11); Genome Biol. 2008; 9(5): R91 (CC BY 2.0)[2] (p 11); TimVickers (CC BY-SA 3.0)[1] (p 13); Lothar Schermelleh (CC BY-SA 3.0)[1] (p 13); Two images Microrao/Public Domain (p 14); USDE/Public Domain (p 14); Karel Schmeidberger (CC BY-SA-3.0)[1] (p 15); Grzegorz Polak (CC BY-SA-3.0)[1] (p 15); Vossman (CC BY-SA 3.0)[1] (p 19); AzaToth/Public Domain (p 19); Image licensed by Ingram Image (p 22); Vaccinationist/Public Domain (p 22); Darekk2 (CC BY-SA 3.0)[1] (p 32); Kelvinsong (CC BY-SA 3.0)[1] (p 38); Cytoskeleton of Rat2-sm9 cells © University of Basel, imaged by R. Suetterlin, courtesy of C-A. Schoenenberger (p 38); Lordjuppiter (CC BY-SA 3.0)[1] (p 39); Four images Roy van Heesbeen/Public Domain (p 41) The Center for Cell Dynamics, University of Washington (p 41); Two images Geoff Morgan (p 44); David Goodsell (CC BY 3.0)[3] (p 45); Egelberg (CC BY-SA 3.0)[1] (p 45); Fourteen images © Geoff Morgan (pp 46–50); Chapman, Frank M. (CC BY 2.0)[2] (p 51); Guide to the coastal marine fishes of California in STATE OF CALIFORNIA, THE RESOURCES AGENCY, DEPARTMENT OF FISH AND GAME, FISH BULLETIN 157 by Miller, Daniel J., creator, Lea, Robert N. (1972). Courtesy of Scripps Institution of Oceanography (p 51); Two images © Geoff Morgan (p 52); Centre for Disease Control/Public Domain (p 52); Johann Georg Sturm, 1796/Public Domain (p 53); Zeynep F. Altun, Editor of www.wormatlas.org (CC BY-SA 2.5)[4] (p 53); Mr.checker (CC BY-SA 3.0)[1] (p 53); Adam Amsterdam (CC BY 2.5)[8] (p 53); Three images © Geoff Morgan (p 54); Oliver Herold (CC BY 3.0)[3] (p 55); Katja Schulz (CC BY 2.0)[2] (p 55); liz west (CC BY 2.0)[2] (p 55); S. Rae (CC BY 2.0)[2] (p 55); SoulRiser (CC BY-SA 2.0)[5] (p 55); prilfish (CC BY 2.0)[2] (p 55); Rushen (CC BY-SA 2.0)[5] (p 55); CSIRO (CC BY 3.0)[3] (p 55); prilfish (CC BY 2.0)[2] (p 55); Yankech gary (CC BY-ND 2.0)[6] (p 55); Seven images © Geoff Morgan (pp 57 –59); Two images licensed by Ingram Image (p 63); Mark Medcalf (CC BY 2.0)[2] (p 63); John Tenniel/Public Domain (p 64); Doug Beckers (CC BY-SA 2.0)[5] (p 64); Bernard DUPONT (CC BY-SA 2.0)[5] (p 65); André Karwath (CC BY-SA 2.5)[4] (p 65); Janice Harney Carr, Center for Disease Control/Public Domain (p 65); Miroslav Duchacek (CC BY-SA 3.0)[1] (p 66); Drc406 (CC BY-SA 3.0)[1] (p 67); Image licensed by Ingram Image (p 67); Paul Asman and Jill Lenoble (CC BY 2.0)[2] (p 67); Image taken from Bolzer et al., (2005) 'Three-Dimensional Maps of All Chromosomes in Human Male Fibroblast Nuclei and Prometaphase Rosettes.' Public Library of Science Journal: Biology 3(5): e157 DOI: 10.1371/journal.pbio.0030157, Figure 7a © Public Library (p 72); Sbharris (CC BY-SA 3.0)[1] (p 73); Two images © Geoff Morgan (p 74); Visual by http://www.pdimages.com (Public Domain) (p 74); Geoff Morgan (p 75); Otis Historical Archives of "National Museum of Health & Medicine" (CC BY 2.0)[2] (p 75); Seven images © Geoff Morgan (pp 75–77); Vuela sobre Moscu/Public Domain (p 79); Geoff Morgan (p 80); GreenBlueRed/Public Domain (p 82); John Tann (CC BY 2.0)[2] (p 82); CDC/Public Domain (p 82); SuSanA Secretariat (CC BY 2.0)[2] (p 82); Two images NIAID (CC BY 2.0)[2] (p 82); History of Medicine/Public Domain (p 84); The poster 'Malaria Kills A Child Every 45 Seconds' reproduced by permission of Christian Aid (p 89); The poster 'Her Love Can't Make Dirty Water Clean' reproduced by permission of Oxfam (p 89); Logo © BioMed Central/The primary paper 'In vivo genome-wide profiling reveals a tissue-specific role for 5-formylcytosine' by Iurlaro M., McInroy G.R., Burgess H.E., Dean W., Raiber E.A., Bachman M., Beraldi D., Balasubramanian S., Reik W. taken from 'Genome Biology' 2016 17:141 (CC BY 4.0)[7] (p 92); Logo © International Journal of Molecular Sciences/ Review article 'DNA Damage: A Main Determinant of Vascular Aging' by Bautista-Niño K., Portilla-Fernandez E., Vaughan D.E., Danser H.J. and Roks, A.J.M. taken from 'International Journal of Molecular Sciences' 2016, 17, 748. (CC BY 4.0)[7] (p 92); Article and figures adapted from: Lalor M.K., Floyd S., Gorak-Stolinska P., Weir R.E., Blitz R., Branson K., et al. (2011) BCG Vaccination: A Role for Vitamin D? PLoS ONE 6(1): e16709. doi:10.1371/journal.pone.0016709) (CC BY 4.0)[7] (p 103); Geoff Morgan (p 106); Geoff Morgan [photograph of pp. 268-269 of the book Colour Identification Guide to Moths of the British Isles by B. Skinner, 3rd ed., (2009 ISBN 978-87-88757-90-3), published by Apollo Books www.apollobooks.com] (p 107).

[1] (CC BY-SA 3.0) https://creativecommons.org/licenses/by-sa/3.0/
[2] (CC BY 2.0) http://creativecommons.org/licenses/by/2.0/
[3] (CC BY 3.0) http://creativecommons.org/licenses/by/3.0/
[4] (CC BY-SA 2.5) http://creativecommons.org/licenses/by-sa/2.5/
[5] (CC BY-SA 2.0) https://creativecommons.org/licenses/by-sa/2.0/
[6] (CC BY-ND 2.0) https://creativecommons.org/licenses/by-nd/2.0/
[7] (CC BY 4.0) https://creativecommons.org/licenses/by/4.0/
[8] (CC BY 2.5) http://creativecommons.org/licenses/by/2.5/

Printed and bound in the UK by Martins the Printers.

CONTENTS

INTRODUCTION

CELLS AND PROTEINS

ORGANISMS AND EVOLUTION

INVESTIGATIVE BIOLOGY

APPENDICES

INTRODUCTION

AN OVERVIEW

USING THIS BOOK FOR REVISION

In this book we have tried to present the course content in a concise form, so that you can follow the ideas and develop your own understanding. It is, however, only an **aid to your revision**. Effective revision has to be an **active process**, so your brain needs to be doing some serious work! This means that you have to:

1. take in information and think about how you can **understand it or explain it** for yourself

2. create your own revision notes using **your own words or pictures**

3. keep **returning** to your revision notes to **reinforce your learning**.

This book has lots of features to help you in your active approach to improving your understanding and your learning:

- Some words or phrases are in **bold** to emphasise their importance. These are key terms and you should be able to explain them, or be able to use them to explain other ideas.

- **Don't forget** items point out key details that might get lost in the big picture.

- We have included **online** items and **video links** which we think will help to reinforce the ideas or explain them in a different way that will help you.

- There are short **online tests** available for each topic at www.brightredbooks.net.

- **Things to do and think about** are sections that take the content of a particular page further, to help you broaden your knowledge or understanding.

ONLINE

The Open University has useful tips on how to revise. Follow the link at www.brightredbooks.net

VIDEO LINK

Find out about the scientifically proven tips for effective studying by watching the clip at www.brightredbooks.net

DON'T FORGET

Keep returning to your revision notes and try to rewrite brief versions from memory to check your learning.

COURSE STRUCTURE

The AH Biology course has three taught units and an individual practical investigation.

Advanced Higher Biology		
Cells and Proteins This unit focuses on the key role that proteins play in the structure and functioning of cells and organisms. The study of protein is primarily a lab-based activity so the unit begins with a selection of important laboratory techniques. The unit then looks at the importance of protein structure and binding to other molecules. Conformational change of proteins is central to controlling their activity and this is studied in the context of membrane proteins, signalling between cells and in the control of cell division.	**Organisms and Evolution** This unit explores the importance of parasites in evolution. After we have introduced field techniques used for investigating evolution in the field, the unit looks at factors influencing the rate of evolution. The maintenance of biological variation necessary to continue evolving in the 'arms race' between hosts and parasites. Sexual reproduction is central to this so we look at sex determination and mate choice. Finally, the Unit considers parasite–host interactions from a range of perspectives.	**Investigative Biology** This unit is half the size of the others and it will give you a solid grounding in the principles and practice of investigative biology in preparation for doing your own practical investigation.
		The **Investigation** provides opportunities for you to put into practice the investigative skills from the taught unit. You will need a good level of self-motivation and organisation as you develop a plan and move on to the collection and analysis of information obtained.

ONLINE

The 'syllabus' can be found at **Course and Unit Support Notes** on the SQA AH Biology website. The **mandatory** and **exemplification** columns list the content of each unit. Follow the link at www.brightredbooks.net

INTERNAL ASSESSMENTS

All three taught units are assessed by your teacher or lecturer. The first part of the assessment for each unit will look at your knowledge, most likely by way of a **written test**, which you must pass. Your school or college will also collect evidence during the course to show that you have developed the necessary **skills of scientific enquiry** at Advanced Higher level; these may also form part of the written tests.

To gain a full pass in the Investigative Biology unit, you must also keep a full record of the **pilot study** for your investigation; this pilot study also allows you to gain full passes in the two larger units. Alternatively, you could do a **practical report** on an experiment to can gain a full pass in the two larger units – but in this case you would still need to complete the pilot study to pass the Investigative Biology unit.

DON'T FORGET

The marks from the unit assessments do not go towards your final exam mark.

HOW THE COURSE IS GRADED

To gain a grade, you must pass the internal assessments for all three units, sit the **written exam** (90 marks) and submit an **investigation report** (30 marks). Your final grade is based on these 120 marks and the grades are A, B, C or D. No grade is awarded if the marks are below those needed for a D grade.

The written exam

This lasts 2 hours 30 minutes and has a maximum possible 90 marks. These will be distributed approximately proportionally across the three taught units. The majority of the marks (about 65) will be awarded for **knowledge and understanding** while the other marks (about 25) will be awarded for applying **scientific inquiry and problem-solving skills**.

selecting information

presenting information

processing information

scientific inquiry and problem solving skills

planning experimental procedures

making predictions

evaluating experimental procedures

drawing conclusions

The exam has two sections:

1. Section 1 has 25 multiple-choice questions, worth one mark each.

2. Section 2 has 65 marks. There will be one **short essay** (4 or 5 marks), one **long essay** (8–10 marks) and a **data-handling** question (7–10 marks). The remaining marks will be awarded to **short-answer** questions.

The investigation report

This must be a piece of individual work and your teacher/lecturer will help you to choose a suitable topic which interests you. You will research the underlying biology and carry out an open-ended investigation involving a significant amount of work that you will carry out without close supervision. For the most part, you will be working autonomously, though your teacher/lecturer will be available to provide support and advice.

The final report is written up in a formal style and will be sent to the SQA for marking. The mark allocation for each section of the report is shown in the table. There are also two marks for aspects of presentation.

Section	Marks
Abstract	1
Introduction	5
Procedures	9
Results	6
Discussion	7
Presentation	2

ONLINE

More information about **the scientific inquiry and problem-solving skills** can be found on page 6–9 of the **Course and Unit Support Notes** on the SQA AH Biology website, a direct link can be found at www.brightredbooks.net

DON'T FORGET

Instead of 'essay', the SQA use the term '**extended-response question**'!

ONLINE

Get to grips with the exam format by looking at recent exam papers at www.brightredbooks.net

 ## THINGS TO DO AND THINK ABOUT

Broaden your interest in biology. Watch out for TV programmes on biological topics, and *NewScientist* regularly has interesting articles about cutting-edge biology. Follow biology-related Twitter accounts for up-to-the-minute news and stories.

SELECTING, PRESENTING AND PROCESSING DATA

As you progress through the Advanced Higher Biology course, you should acquire the skills to deal with complex data. Try to experience laboratory or field biology so that you become comfortable with the challenges of dealing with the variability of real data.

Below is some background to data-handling skills and some guidance on what you may be expected to do during the course. In the end, you will be expected to be able to show off a selection of these skills in your investigation, unit assessments or examination.

SELECTING INFORMATION

Data

Discrete and continuous data. Discrete data are those that naturally fall into distinct categories. Continuous data are those that vary from one extreme to another and, as a result, can be measured on a scale. Box plots show the spread of the data.

Sampled data. It is usually not possible to measure or observe every member of a population. Instead, a sample of data is collected. To be reliable, it is important to select a sample that is representative of the variation in the population as a whole – i.e. it is important that a sample is large enough to allow reliable results.

Charts

You must be able to select data from identification keys, tables, pie and bar charts, line graphs, lines of best fit, graphs with semi-logarithmic scales, graphs with error bars and information presented as box plots. Logarithmic scales increase in an exponential rather than linear manner. Each tick-mark is 10 times higher than the previous one.

Analysis

You are expected to be able to select information with reference to **three sources** at the same time. For example, you may have to find a relevant piece of information in text and use that to locate information in a table, and then use that to interpret a graph!

Statistics

Standard deviation and error bars. Standard deviation is a measure of the variability of the data. If this variability is plotted as an error bar, it allows you to judge the spread of the data that were combined in the average. Data that is very reliable has a smaller standard deviation than unreliable data.

Significance. This is the scientific measure of the likelihood or probability of a result occurring by chance. Significantly different results are difficult to explain by chance alone.

PRESENTING INFORMATION

Tables

Tabulating data is a key scientific skill and is assessed in the investigation report. Remember that the first column of data is usually the **independent variable**, and the remaining columns are raw **dependent** data and data derived from calculations.

Charts and graphs

You should be very familiar with bar charts and pie charts. These are used for discrete, rather than continuous, data.

Remember that the independent variable is plotted along the x-axis and the dependent variable is plotted up the y-axis. Make sure that the y-axis on a bar chart has a proper scale and label 0 (zero) at the origin if appropriate.

The two types of graph that you should be able to draw are the line graph and the histogram. These are both used for continuous data. The line graph is used to visualise correlations between two continuous variables. Remember that the independent variable

DON'T FORGET

Quantitative data refers to numerical results and data that are the result of making measurements. Qualitative data refers to descriptive observations.

contd

is plotted along the x-axis and the dependent variable is plotted up the y-axis. Make sure that you use proper scales along these axes and label 0 (zero) at the origin, if appropriate, on each one. For a histogram, the categories on the x-axis are convenient divisions of a continuous variable and the bars are touching.

Communication

Written communication. Science is pointless if we do not communicate the findings, so make sure that you develop great communication skills during this course. Your scientific communication is assessed in your pilot study write-up (or practical report), in your investigation and in the examination. Write clearly, coherently and logically.

Oral communication. Scientists communicate verbally at seminars, conferences and in the (coffee) bar, so make sure that you spend time in class verbalising your ideas and double-checking that you know the right vocabulary and how to use it.

PROCESSING INFORMATION

Number

You will need to be able to perform calculations using numbers in order to process data. These could involve whole numbers, decimals or fractions. In general, there are no marks for working (even when you are given space for working).

Proportion

At the heart of any science is the need to compare data, so it is important that you understand the concept of proportion and the associated mathematical tools. To compare different numbers, we use a ratio or a fraction. To compare different fractions more easily, we use percentages. A common requirement is to calculate a percentage change so remember:

$$\text{percentage change} = [(\text{final} - \text{original})/\text{original}] \times 100$$

Accuracy

It is important to consider accuracy when processing data. When calculating averages, do not express the answer to many more decimal places than the original data. Do not go to more than one more decimal place than the original data.

Units

Units should always be stated and should be m, s, ml, kg, g, etc. Do not leave units out. Do not use imperial units. It is fine to express units using a solidus (e.g. m/s) or as scientific notation (e.g. ms^{-1}).

Remember that there are 1000 nano (n) in one micro (μ). There are 1000 μ in a milli (m). There are 1000 m in a unit. One thousand units is a k. One thousand k is mega (M), and 1000 M is giga (G).

Means and ranges

You have to be able to work out a mean average and the range of a set of data. You will not have to calculate a standard deviation in the exam, though it might be very useful in your investigation. Use spreadsheets to help you process data.

 THINGS TO DO AND THINK ABOUT

Biologists can be wrong ... sometimes. The range is properly defined as the number of units between the lowest and highest value. However, biologists usually state the **range** as the lowest to the highest value; this is really a measure of the **spread** of the data.

 **DON'T FORGET**

Remember that in bar charts there should be spaces between the bars – just like in jail!

 **DON'T FORGET**

In a histogram, plot the bars next to one another without any gaps – a histogram is a squishtogram!

 **DON'T FORGET**

There are no half marks in Biology.

 **DON'T FORGET**

If there has been a decrease, then the percentage change figure will be negative. Alternatively, it can be called a percentage decrease and then the negative symbol is not required.

LABORATORY TECHNIQUES FOR BIOLOGISTS: LIQUIDS AND SOLUTIONS

DILUTION SERIES

The study of cells and proteins in the laboratory requires careful development of basic laboratory skills. Many substances are found in cells at extremely low concentrations. To recreate cellular conditions in the laboratory requires an understanding of **dilution** and **measurement uncertainty**. To ensure accuracy, it is important to use appropriate measuring methods whether using scales, measuring cylinders, pipettes or autopipettes.

Dilutions are used in many experimental procedures. They can be used to control potential confounding variables in an experimental system, to generate a suitable range in an independent variable or as a way of modifying the dependent variable, so that a measureable value can be obtained.

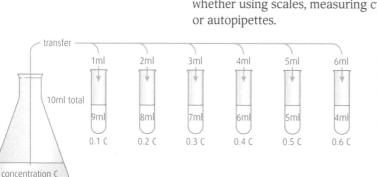

A linear dilution series that would make six dilutions, each with a volume of 10 ml. To calculate the volume of stock solution required (V_1) to make a new concentration (C_2) the formula $V_1C_1 = V_2C_2$ can be used. Note that in this example V_2 is always 10 ml.

Linear dilution series

A linear dilution series consists of a range of dilutions that differ by **an equal interval**. For example, solutions of concentrations 0·1, 0·2, 0·3, 0·4, 0·5 and 0·6 ml would represent a linear dilution series.

To make a linear dilution series, it is normal practice to add different volumes of stock solution to different volumes of solvent. In this way, **each concentration is made individually** and any measurement errors affect only the one concentration.

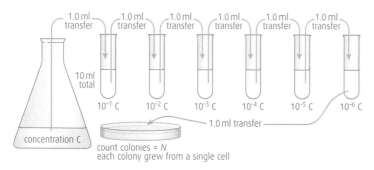

A log dilution series involving six 10-fold dilutions. Note that each dilution acts as the stock to make the next dilution in the series. Log dilution series are often used to allow the estimate of microbial cell density. In the case at the top of this page, the number of colonies is countable on the 10^{-4} and 10^{-5} plates which allows the undiluted concentration of viable cells (C) in the original culture to be calculated. C = approximately $2·6 \times 10^7$ cells per ml.

Log dilution series

A log dilution series consists of a range of different dilutions that differ by a **constant proportion**. For example, solutions of concentrations 10^{-1}, 10^{-2}, 10^{-3}, 10^{-4}, 10^{-5} and 10^{-6} would represent a log dilution series.

To make a log dilution series, it is normal for each dilution solution to act as the stock for the subsequent dilution. In this way, each concentration depends on those made before and any earlier measurement **errors are compounded** in later dilutions.

DON'T FORGET

Any measurement errors made while making up any stock solution will be compounded if further dilutions are made.

COLORIMETER

A colorimeter is used to measure the **concentration of a pigment** in a solution, the **turbidity** of liquid or the **density of cells** in a culture. It does this by illuminating a small sample of the test substance, held in a small transparent cuvette, using a coloured light source and electronically recording how much of the light is **absorbed** by the sample. For turbidity, a denser sample will show a lower degree of **transmission**. For each experiment

contd

the machine is **calibrated** using a 'blank' cuvette containing solvent only, which acts as a baseline or control value for the comparison.

Typical uses of a colorimeter in schools include measuring:

- the density of a cell culture growing in a liquid broth – careful aseptic technique is required for this
- the concentration of betanin leaking from beetroot cells whose membrane proteins have been denatured by heat
- the rate of reaction of enzymes such as dopa oxidase
- and quantifying the result of an ELISA demonstration.

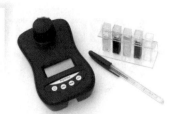
A colorimeter along with a series of cuvettes. Do not expect your colorimetry investigation to be as colourful as this!

STANDARD CURVE FOR DETERMINING AN UNKNOWN

A standard curve is made by plotting the absorbance readings for a series of **known concentrations** of a substance or culture. Once the line (the standard curve) has been produced, it can be used as a reference for any samples of unknown concentrations of the same substance or culture. Through **interpolation**, the concentration of the unknown can be estimated.

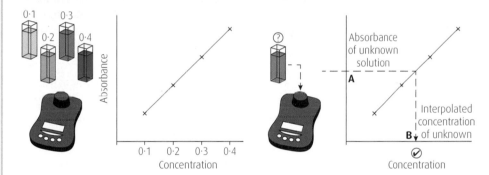

The absorbance of standard solutions of known concentration is measured in a colorimeter – in this case, 0·1, 0·2, 0·3 and 0·4 units. The standard curve is plotted using these data. The standard curve can now be used to determine the concentration of an unknown, once its absorbance has been measured. It is only possible to interpolate from a standard curve, so if the unknown appears to have a concentration outside the initial range of the standards, then a greater range of standards must be measured.

BUFFERS TO CONTROL PH

Buffers are aqueous solutions that show **very little variation in their pH** despite addition of acids or alkalis. As almost all biological processes are pH dependent, cells and their secretions tend to contain pH buffers. For example, in mammalian blood the presence of carbonic acid and bicarbonate anions buffers the blood pH and keeps it between the values of pH 7·35 and 7·45. Any variation beyond either pH 6·8 or 7·8 is likely to be lethal.

In laboratory experiments, buffers can be selected so that the pH of solutions can be controlled. In cell-culture media, buffers are used to prevent pH changes that could otherwise occur as a result of the build-up of waste products.

 THINGS TO DO AND THINK ABOUT

Health and safety

Many chemicals or organisms can be intrinsically hazardous and their use in the laboratory may involve risks to people, to other organisms or to the environment. Before starting any practical work, it is essential that **hazards are identified**. Risk is the chance of harm occurring through exposure to these hazards. Control measures are put in place to **reduce the risk** to an acceptable level. The control measures of last resort include the wearing of personal protective equipment such as gloves, goggles and lab coat. It is worthwhile developing the skills of risk assessment during this course.

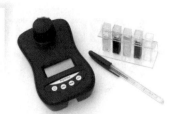
VIDEO LINK

Watch the video at www.brightredbooks.net to see a portable colorimeter in action.

VIDEO LINK

Head to www.brightredbooks.net and watch the clip for a demonstration of how to determine the molecular weight of an unknown protein after gel electrophoresis.

DON'T FORGET

Interpolate, don't extrapolate!

ONLINE

There is a useful overview of risk assessment at www.brightredbooks.net

ONLINE TEST

Head to www.brightredbooks.net to test yourself on this topic.

LABORATORY TECHNIQUES FOR BIOLOGISTS: SEPARATION TECHNIQUES

CENTRIFUGATION

Centrifugation, along with other separation techniques, is a vital tool in the study of cell biology. The enormous complexity of cell biology cannot be studied easily in whole cells. The rapid advance in scientific understanding of cells and proteins has been due to a reductionist approach whereby the various constituent cellular structures and molecules are separated and studied in isolation. Once their basic roles and functions are understood, a systems approach can be taken and the knowledge can be assembled into a coherent whole.

Centrifugation is a method for separating materials **in suspension** according to their **density**. Since almost everything within living cells is essentially in suspension, this is a suitable method for separating different cell constituents. Biological fluids require little preparation, but solid tissue is homogenised in a blender first with the addition of buffer. The material is rotated in a **centrifuge tube** at between **2000 to 120 000 revolutions per minute**. The resultant **g**-force causes the constituents to separate. The most dense items form a **pellet** at the bottom of the tube, whereas the liquid fraction above is termed the **supernatant**.

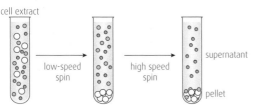

Careful consideration of centrifugation speed and duration results in different fractions in the pellet.

Typical uses of a centrifuge in a school setting include enzyme extractions from tissue, such as potato-starch phosphorylase; the use of centrifugation ensures that the enzyme in the supernatant is separated from any starch, which ends up in the pellet for disposal. In the study of the activity of chloroplast electron-transport chains (also known as the Hill reaction), a centrifuge can be used to separate the photosynthetic membranes from other cell constituents; in this case the required fraction is the pellet and the supernatant is discarded.

CHROMATOGRAPHY

Paper and thin-layer chromatography

Thin-layer chromatography separation of photosynthetic pigments from green plants. This result demonstrates that there are several different light-harvesting membrane pigments within the chloroplast.

In paper and thin-layer chromatography, amino acids can be separated according to their characteristics of **solubility**. The materials being tested are spotted toward the base of the chromatography medium and a solvent mix pulls the different constituents up the chromatogram. The different **relative affinities** of the materials being tested for the medium results in different rates of movement, so the constituents travel to different heights.

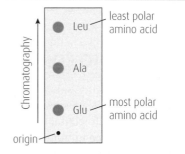

In the chromatography of amino acids, the chemical properties of the R-group determine the distance travelled up the chromatogram.

In paper chromatography, the cellulose paper is highly hydrophilic and the solvent used has hydrophobic properties. Similar solvent properties are used in thin-layer chromatography but the thin layer is made from silica gel or cellulose.

Affinity chromatography

Affinity chromatography is a technique for the separation of one specific protein from a mixture. An **antibody or ligand** specific for binding with the protein in question is immobilised in an agarose gel packed into a column. The mixture of proteins is poured through the column. Only the protein of interest binds to the antibody or ligand within the column. The other proteins will not bind and can be washed away. By then **washing**

ONLINE

A thorough description of affinity chromatography is available at www.brightredbooks.net

contd

the column with either a buffer of different pH (which lowers the affinity of the protein to the ligand) or with a solution containing free ligand, the target protein can be recovered.

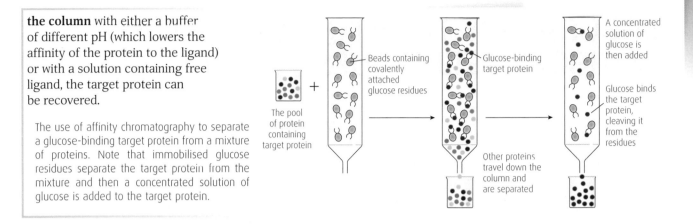

The use of affinity chromatography to separate a glucose-binding target protein from a mixture of proteins. Note that immobilised glucose residues separate the target protein from the mixture and then a concentrated solution of glucose is added to the target protein.

The pool of protein containing target protein

Beads containing covalently attached glucose residues

Glucose-binding target protein

Other proteins travel down the column and are separated

A concentrated solution of glucose is then added

Glucose binds the target protein, cleaving it from the residues

PROTEIN ELECTROPHORESIS

Protein electrophoresis uses current flowing through a buffer to separate proteins. **Size** and **charge** are factors that affect the rate at which any particular protein migrates through a gel. It is possible to run gels with the proteins in their native non-denatured (folded) state or in a non-native denatured (unfolded) state. To unfold the protein, it is denatured by heat in the presence of detergent – the result is an unfolded linear protein with uniform charge along its length; its rate of migration now depends only on its size.

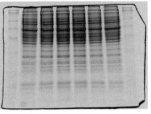

An electrophoresis gel of denatured muscle proteins stained with Coomasie blue so that each protein band can be visualised. The outermost lanes contain a kaleidoscope standard; these standards allow a standard curve to be plotted and the molecular weight of the other proteins to be estimated.

ISOELECTRIC POINT

The isoelectric point of a protein is the pH at which it has an **overall neutral charge**. Above and below this pH there will either be a majority of positive or negative charges at the surface of the protein and the protein will be soluble. Water molecules interact with these charges, keeping the protein suspended in solution. At the isoelectric point, the overall neutral charge allows the protein to form a solid and **precipitate** out of solution. This technique can be used to separate proteins from one another. By slowly changing the pH of a solution, successive proteins can be separated. For example, casein in milk has an isoelectric point of pH 4.7 and will precipitate at this point. The curdling of milk in cheese manufacture occurs at this pH.

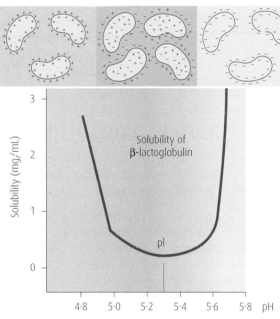

The isoelectric point (pI) of beta-lactoglobulin is 5·3 as this is the pH at which it has an overall neutral charge. Under these conditions the protein forms larger aggregations to precipitate.

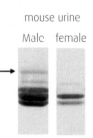

mouse urine

Male   female

This study used non-denatured proteins to show that the gender of a mouse can be determined through the presence or absence of a particular protein in its urine.

VIDEO LINK

Watch a clear demonstration of the electrophoresis of fish muscle tissue at www.brightredbooks.net

THINGS TO DO AND THINK ABOUT

An interesting investigation could involve the following steps. The primary sequence of a protein of interest could be found by searching an online database. Using the website isoelectric.ovh.org, the isoelectric point of the protein could be predicted. In the laboratory, the isoelectric point of the protein could be determined experimentally.

ONLINE TEST

Test yourself on this topic online at www.brightredbooks.net

LABORATORY TECHNIQUES FOR BIOLOGISTS: ANTIBODY TECHNIQUES

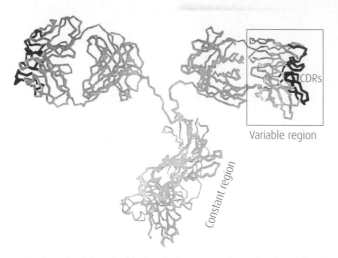

Variable region

CDRs

Constant region

Each antibody is a double-headed structure. The ends of each head are variable – this means that they differ from one specific antibody clone to another. All the antibodies produced by any one B lymphocyte are, of course, absolutely identical in structure. At the tip of the variable region, coloured in red here, is the unique arrangement of polypeptide R-groups involved in binding with the antigen. These areas are known as the complementarity determining regions.

ANTIBODIES

Antibody techniques are powerful tools in laboratory studies for the **detection and identification of specific proteins**. The ability of antibodies to bind to specific antigens is utilised in antibody techniques.

Antibodies are Y-shaped globular proteins produced by B lymphocytes as part of the immune response of a vertebrate. Each B lymphocyte produces one specific antibody that **binds** to one specific antigen. This binding helps to render an antigen harmless, as described on page 87. The spleen and bone marrow are the sites of production of B lymphocytes.

A serum containing antibodies can be harvested from the blood of animals that have been exposed to a particular antigenic material. Some time after exposure, blood is removed and antibodies are separated by centrifugation. Many different antibodies will have formed to different parts of the antigen. Each antibody is made by a single B lymphocyte clone. A serum made in this way with many different antibodies against an antigen is described as **polyclonal**.

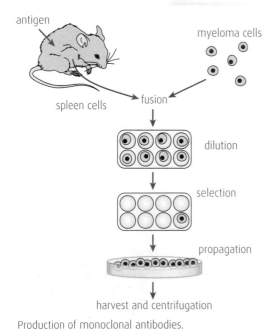

antigen

myeloma cells

spleen cells

fusion

dilution

selection

propagation

harvest and centrifugation

Production of monoclonal antibodies.

MONOCLONAL ANTIBODIES

The generation of monoclonal antibodies – a supply of antibodies that are **identical** and will **bind to exactly the same feature of the antigen** – is an essential stage in all antibody techniques.

To produce pure monoclonal antibodies, a single line of **B lymphocytes** must be grown, each secreting the same specific antibody. B lymphocytes can be harvested from the **spleen** of a mouse that has been exposed to a specific antigen several weeks earlier.

B lymphocytes do not divide in culture. To get around this problem, the B lymphocytes are **fused** with **myeloma** (cancer) cells from an immortal cell line, using **polyethylene glycol** (PEG). The cells produced are called **hybridomas**. Dilution ensures that each hybridoma is placed in its own screening well. Through the use of selective media, the hybridoma cell line capable of producing the specific antibody of interest can be identified. Cultures can then be grown in fermenters and each immortal hybridoma cell line produces only one type of antibody (so it is monoclonal). The secreted antibody is extracted and purified using centrifugation.

IMMUNOASSAY TECHNIQUES USING REPORTER ENZYMES

Monoclonal antibodies are used in the **diagnosis and detection of disease**. For example, **immunoassay techniques** (such as ELISA) involve the use of monoclonal antibodies that have an enzyme attached. This **reporter enzyme** catalyses a colour-change reaction

contd

that is used to detect and quantify the presence of a specific antigen. If the antigen is present, the antibody binds allowing the reporter enzyme to produce a coloured product – visual confirmation of the antigen's presence. If the antigen is not present, the antibody cannot bind and the enzyme is washed away before any reaction can take place. Rather than manufacture a specific antibody with a linked reporter enzyme for every possible antigen of interest, two antibodies are usually used. The first is specific to the antigen and the second is specific to the constant region of the first type of antibody. Blood samples can be screened for either antigens or antibodies for particular pathogens, such as HIV or meningitis. A laboratory demonstration for ELISA involves the detection of the antigens from the fungus *Botrytis*, a pathogen of soft fruit. Pregnancy testing kits use the immunoassay technology to detect the presence in the urine of human chorionic gonadotrophin (HCG), a hormone only produced by the placenta.

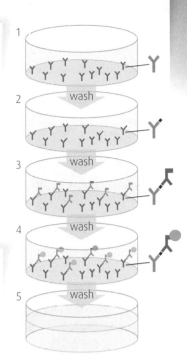

FLUORESCENT-LABELED PROTEIN BLOTTING

Fluorescent-labeled protein blotting is a technique for **identifying specific proteins** that have been separated using gel electrophoresis. The proteins are **blotted** from the gel onto a nitrocellulose membrane or nylon filter, recording their final positions on a more convenient material.

The filter is then flooded with **fluorescent-labeled monoclonal antibodies**. Several different antibodies can be used simultaneously, as long as each has a differently coloured fluorescent marker. Once the antibodies have bound to their target proteins, the excess is washed away. When exposed to light of particular wavelengths, the fluorescent-labeled antibodies allow the precise location of their specific target proteins to be identified. A fluorescent-labeled protein blot allows the location of the target protein to be precisely identified.

Size (kDa)

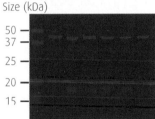

A fluorescent-labeled protein blot allows the location of the target protein to be precisely identified.

This ELISA involves an antibody (1) attached to the bottom of a plastic well. If a test sample contains the specific antigen then this will bind (2) to the antibody. If an enzyme-linked antibody is then added (3), it will bind only where the antigen has bound. If the enzyme substrate is added (4), then a colour-change reaction occurs (5). It is vital that the test well is thoroughly rinsed between each stage to prevent false-positive results being obtained.

FLUORESCENT-IMMUNOHISTOCHEMICAL STAINING

Histochemistry is the study of tissues using stains and microscopy. Fluorescent-immunohistochemical staining is a technique that is used to **visualise the distribution of specific cellular components in live cells**. In fluorescence microscopy particular protein structures can be visualised in a way that previously was not possible.

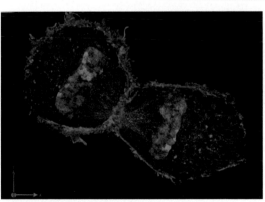

Fluorescent-immunochemical staining of mouse cell telophase. The nucleosomes are fluorescing blue, the tubulin of the spindle fibres fluorescing orange and the actin cytoskeleton is fluorescing green. In each case, a specific fluorescent-labeled antibody has bound to each type of protein to show the precise location of each subcellular structure.

⊕ DON'T FORGET

An assay determines the presence or concentration of a substance in a mixture; an immunoassay uses a monoclonal antibody to do this.

▶ VIDEO LINK

Watch the description of the use of fluorescent-labeled protein blotting in research into neurodegenerative disease in Scotland at www.brightredbooks.net

✓ ONLINE TEST

Test yourself on this topic at www.brightredbooks.net

💭 THINGS TO DO AND THINK ABOUT

Monoclonal antibodies are increasingly being used in the treatment of disease. In this technology, toxins are attached to tumour-specific monoclonal antibodies. When the antibodies reach the tumour cells, they combine with their specific antigens, bringing the toxins into close contact with the target cells and killing them. If you want to find out more, research the use of the monoclonal antibody trastuzumab in cancer treatment.

LABORATORY TECHNIQUES FOR BIOLOGISTS: ASEPTIC TECHNIQUES AND CELL CULTURE

(a)

(b)

(c)

CELL CULTURE

Cell and tissue culture is the growth of cells, tissues or organs in artificial media in a laboratory. Culturing can produce many **genetically identical clones** of an initial cell sample.

Some cell types can be grown in suspension using liquid medium. Other types can be grown on, or within, a solid medium (substrate) such as agar jelly. To ensure rapid growth of the chosen cells, optimum conditions are provided in terms of nutrients, pH and gases. Anaerobic organisms can be cultured in non-aerated suspension or within agar.

The initial sample can be added in several different ways. In an **inoculum**, growing cells are added in a volume of liquid medium from a previous culture, such as in the setting up of bacterial fermenters. In some animal-cell culture applications, individual **cells** – removed enzymatically from tissues – may be added to the culture. Alternatively, **explants**, which are small cuttings of whole tissue, may be added to the growth medium; this is particularly successful in plant-tissue culture.

Different organisms require different media to grow well in the laboratory. Here single colonies are isolated using streak plates: (a) *Micrococcus luteus* on nutrient agar; (b) *Vibrio cholera* on bile salt agar; (c) *Escherichia coli* on MacConkey agar.

ASEPTIC TECHNIQUE

Contamination is the enemy of cell or tissue culture. Aseptic technique is a set of **precautions taken to prevent contamination**; the use of sterile materials and the appropriate treatment of the source tissue are vital to prevent accidental inoculation with unwanted cells or spores. Bacterial or fungal contamination will rapidly outcompete and spoil a culture of slower-growing plant or animal cells.

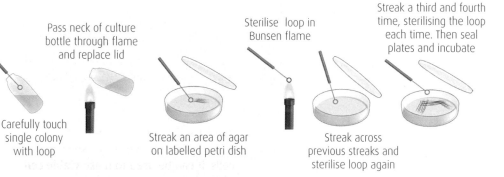

Sterilise loop in Bunsen flame

Pass neck of culture bottle through flame (keep lid in hand)

Pass neck of culture bottle through flame and replace lid

Carefully touch single colony with loop

Sterilise loop in Bunsen flame

Streak an area of agar on labelled petri dish

Streak across previous streaks and sterilise loop again

Streak a third and fourth time, sterilising the loop each time. Then seal plates and incubate

An example of good aseptic technique: the stages in subculturing a microorganism from a pure colony on an agar slope. The streak plate technique is an ideal technique for the detection of contamination and should provide pure colonies for further subculture.

ONLINE TEST ✓

Test yourself on this topic at www.brightredbooks.net

DON'T FORGET ✚

A cell line is a genetically uniform cell culture developed from a single cell.

CELL LINES

Mammalian cells are usually cultured in liquid in a flask. To undergo rapid growth and cell division in culture, most mammalian cells require the addition of a complex medium containing chemical **growth factors**. Growth factors can be provided by the addition of an **animal serum**, such as foetal bovine serum (FBS). FBS is a mixture containing growth factors, proteins, salts, vitamins and glucose. Antibiotics are added to minimise the chances of spoilage by microorganisms.

Cells for culture are detached from source tissue using proteolyic enzymes such as trypsin. When the cells are added to the flask they adhere to the surface of the flask,

contd

spread out and then start to divide. They will stop dividing when they have formed a complete monolayer across the bottom of the flask.

The cells soon use up the nutrients in the medium, so, to keep the cloned **cell line** alive, some must be subcultured into a fresh culture flask. Newly created animal cell lines tend to **die** after a finite number of divisions (about 60 – the Hayflick limit). This reduces the length of time a **primary cell-line** culture can be maintained. However, this limit does not exist in **cancer cell lines**, which are immortal and can be subcultured indefinitely.

PLANT TISSUE CULTURE

Certain plant cells are **totipotent** and so are able to differentiate into all the cell types required to form a whole new organism. Explants are **small pieces of plant tissue** that are placed on a solid medium. **Growth regulators** can be used to induce embryogenesis to generate whole new embryonic plants. Organ formation can also be stimulated by altering the ratio of growth regulators in the medium; for example, cytokinins promote shoot growth, whereas auxins promote root growth.

Animal cell culture can either be (a) in flasks or (b) in suspension in liquid medium. In either case the cells require a complex medium containing growth factors from animal serum. Primary cell lines will have a limited lifespan but cancer cell lines can divide indefinitely, as long as they are subcultured regularly.

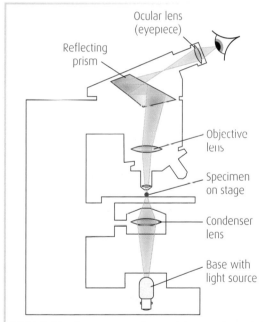

Bright field microscopy can be used to examine whole organisms, parts of organisms or sections of dissected tissue. The specimens must be small and thin enough to mount under a coverslip on slides or cavity slides. Stains are often used to improve the definition. Magnifications up to ×400 are commonly used and, with the use of oil immersion, ×1000 is possible.

USING HAEMOCYTOMETERS

A haemocytometer is a graduated microscope slide used to **count cell density**. It was originally designed to count the density of blood cells, hence its name. The coverslip of a haemocytometer slide is designed to sit a known distance from the slide. This means that the viewer is looking through a **known volume of medium**. By counting the number of cells in a particular area of the grid, a cell density can be calculated.

A haemocytometer can be used to make **total cell counts**, or if vital staining is used to distinguish living cells, it can be used to make **viable cell counts**. An example of a vital stain is methylene blue; when yeast is viewed under the microscope at high power using this stain, the vital or living cells are colourless, whereas dead cells are blue. It is often useful to know what proportion of a cell culture is viable.

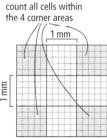

cell suspension

0.1 mm gap

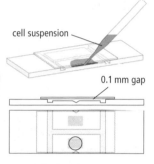

count all cells within the 4 corner areas

1 mm

1 mm

A haemocytometer slide allows the number of cells in a known volume to be counted. If all of the cells within the four corner $1\,mm^2$ squares and the centre $1\,mm^2$ square are counted, this gives the total cell count per $0.5\,mm^3$ of medium (as the depth is given as 0.1 mm). Consistent counting rules must be followed to prevent overestimating abundance. So, cells in contact with only two edges of the squares are considered to be within the square.

THINGS TO DO AND THINK ABOUT

Could a doctor start a cell line using your cells without your permission? Do human cell lines have 46 chromosomes? Why do some cell lines become 'laboratory weeds'? Find out more about the *HeLa* cell line to answer these questions and maybe read the bestselling book *The Immortal Life of Henrietta Lacks*.

VIDEO LINK

Watch the tutorial in the use of haemocytometers to calculate cell density at www.brightredbooks.net

PROTEOMICS AND AMINO ACIDS

PROTEOMICS

An organism's **genome** is its complete set of DNA, including both the protein-coding genes and the non-coding regions of the DNA. The number of protein-coding genes in the human genome is not accurately known and estimates vary between 20 000 to 25 000. All cells will express the important housekeeping genes (such as those for respiration and protein synthesis) but they will only express a sample of the other genes, producing the proteins characteristic for that type of cell.

The **proteome** is the entire set of proteins that can be expressed from a genome and this is much larger than the number of genes. For example, the human proteome is estimated at between 250 000 and a million different proteins. The proteome is much larger because of:

1. **alternative RNA splicing** – depending on which RNA segments are treated as exons and introns in a primary RNA transcript of a gene, different mature mRNA molecules are produced

2. **post-translational modification** – the polypeptide made at the ribosome can be cut and combined in different ways and can also have phosphate or carbohydrate groups added to it.

The proteins that are produced by a cell can change due to the effects of disease. This means that specific marker proteins in the proteome can be early indicators of conditions such as heart disease or cancer.

DON'T FORGET

All cells in an organism have the same set of genes. Cellular differentiation is due to differences in gene expression, RNA splicing and post-translational modification to form the particular proteome of each specific tissue.

ONLINE

Read about how Glasgow University researchers used proteomics to find that raw olive oil in the diet reduces the presence of heart disease biomarkers at www.brightredbooks.net

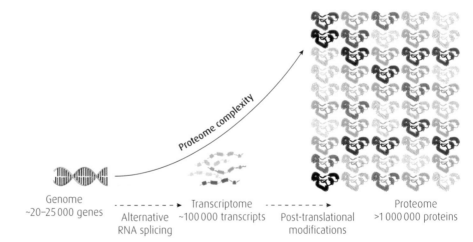

Genome
~20–25 000 genes

Alternative RNA splicing

Transcriptome
~100 000 transcripts

Post-translational modifications

Proteome
>1 000 000 proteins

Proteome complexity

How a genome codes for a much larger proteome.

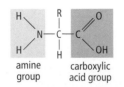

amine group

carboxylic acid group

The structure of amino acids.

AMINO ACIDS

Structure of amino acids

Polypeptides are polymers of amino acid monomers. When these polypeptides undergo post-translational modification they become proteins.

All amino acids have a central carbon atom which has four groups bonded to it: an NH_2 amine group at one side, a COOH carboxylic acid group at the other, and the remaining two bonds are taken by a hydrogen and a variable R group. When in aqueous solution in the cell, the NH_2 group will gain a hydrogen ion to form NH_3^+ while the COOH group will lose a hydrogen ion to form COO^-.

There are four classes of amino acids: **acidic, basic, polar and hydrophobic**. These classes are defined by the **functional group** of the R group and these properties are very important in the structure of proteins.

contd

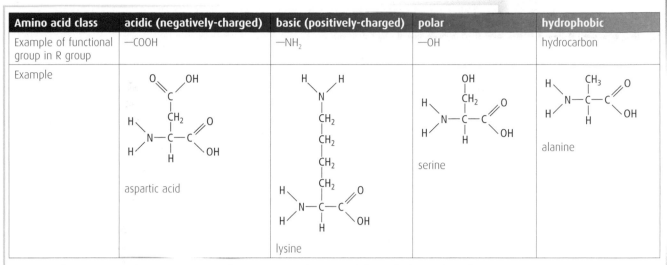

Amino acid class	acidic (negatively-charged)	basic (positively-charged)	polar	hydrophobic
Example of functional group in R group	—COOH	—NH$_2$	—OH	hydrocarbon
Example	aspartic acid	lysine	serine	alanine

Because the R groups also interact with the aqueous solutions of the cell, the carboxylic acid groups will become negatively charged (COO⁻), while the basic NH$_2$ groups will become positively charged (NH$_3^+$). Polar R groups (e.g. OH), have groups which are very slightly charged, so can form hydrogen bonds with other molecules such as water. Hydrophobic R groups carry no charge so do not form any hydrogen bonds with water, hence they do not mix readily with water.

Linking amino acids

Amino acid monomers are linked together during translation at the ribosome. An enzyme causes a **condensation** reaction between two adjacent amino acids. A water molecule is removed by joining the OH of the COOH of one amino acid to a hydrogen from the NH$_2$ of the other amino acid. The bond that links the amino acids is called a **peptide bond**.

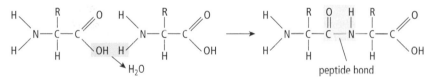

A condensation reaction forming a peptide bond.

Because they are all linked by peptide bonds, a chain of amino acids is called a **polypeptide**. The chain has an NH$_2$ group at the **N-terminus**, and a COOH at the **C-terminus**. The order in which the amino acids are synthesised into a polypeptide chain from N-terminus to C-terminus is called the **primary structure**.

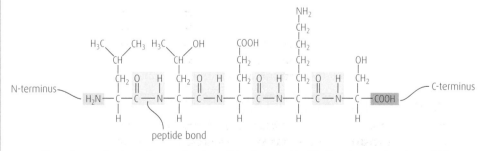

The primary structure of a small polypeptide. Can you identify the classes of amino acids?

THINGS TO DO AND THINK ABOUT

The pH of a protein's environment affects the ionisation of the acidic and basic R groups and, depending on their position, these groups can be easy or difficult to ionise. Solutions of different pH affect the overall charge on a protein and are used as part of the electrophoresis and isoelectric separation techniques.

DON'T FORGET

You do not need to know the structure of any specific amino acids. You should be able to identify the features of the general structure and the functional part of the R group on a diagram.

DON'T FORGET

The amino-acid sequence determines the protein's structure.

VIDEO LINK

Check out the nice introduction to amino acids from the Khan Academy at www.brightredbooks.net

ONLINE TEST

Head to www. brightredbooks.net to test yourself on this topic.

PROTEIN STRUCTURE

PRIMARY STRUCTURE

DON'T FORGET

The primary structure determines all the higher levels of a polypeptide's structure.

The primary structure is the sequence in which the amino acids are synthesised into the polypeptide. This amino-acid sequence determines the protein structure and, hence, the function of the protein.

Primary sequence of anti-diuretic hormone. This small polypeptide travels in the blood and has amino acids that are polar (green) and basic (blue) and which help it to mix with the aqueous plasma.

SECONDARY STRUCTURE

Secondary structure is stabilised by **hydrogen bonds** along the backbone of the polypeptide strand. These hydrogen bonds exist **between different peptide bonds** in the chain: the hydrogen of the N–H has a weak positive charge, so it is electrically attracted to the weak negative charge on the oxygen of the C=O of another peptide bond.

There are three types of secondary structure.

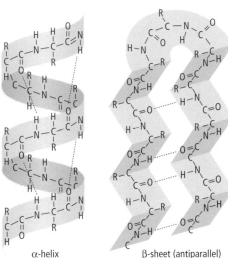

α-helix   β-sheet (antiparallel)

Secondary structure, showing an **α-helix**, and a **β-sheet** with a **turn**. Dotted lines show hydrogen bonds between NH and C=O.

- The **α-helix** is a spiral with the R groups sticking outwards.

- The **β-sheet** has parts of the polypeptide chain running alongside each other to form a corrugated sheet, with the R groups sitting above and below. The **β-sheets** are usually **antiparallel β-sheets** (chains in **opposite directions** with respect to N–C polarity) but they can also be **parallel β-sheets** (chains in the **same direction** with respect to N–C polarity).

- The polypeptide chain can also form **turns** where the chain folds back on itself.

TERTIARY STRUCTURE

The **final folded shape** of the polypeptide is called the tertiary structure. This three-dimensional conformation contains regions of secondary structure which are stabilised in position by **interactions between R groups** of amino acids. The R groups may have been far apart in the primary structure but the folding at the secondary level brings some R groups close enough to interact.

Possible interactions between the R groups are shown in the table below.

Interactions between R groups	Description
hydrophobic interactions	As a polypeptide folds into its functional shape, the hydrophobic R groups are repelled by water and so usually end up to the inside of the polypeptide.
ionic bonds	COOH and NH_2 groups ionise to become COO^- and NH_3^+. These groups are strongly charged, so attract each other.
van der Waals interactions	Very weak attractions between the electron clouds of atoms.

Hydrogen bonds are a particularly strong example of these. |
| disulphide bridges | Covalent bonds form between sulfur-containing R groups of cysteines. |

contd

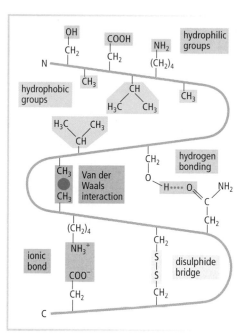

These interactions can be affected by changes in temperature and pH. A higher temperature will cause the tertiary structure to become destabilised, leading to denaturation. The increased heat provides more kinetic energy and so the polypeptide chain shakes more, breaking the weaker ionic bonds, van der Waals interactions and hydrogen bonds. Changes in pH affect the ionisation of the acidic and basic R groups, changing the charge that they carry, so they no longer bond correctly and the polypeptide unfolds. To see these effects on a real protein, try out the activities in *Things to do and think about*.

The interactions between R groups in tertiary structure.

Quaternary structure of collagen protein with three polypeptide chains.

QUATERNARY STRUCTURE

Many proteins have more than one polypeptide subunit which connect by bonding between their R groups. For example, collagen is made of three polypeptide subunits. Some proteins, such as lysozyme or myoglobin, have only one polypeptide chain so the tertiary structure is the final level of the structure.

Proteins with quaternary structure can show cooperativity effects between their polypeptide subunits. Bonding between the subunits means that changes to the conformational shape of one polypeptide subunit can affect the properties of the other subunits.

PROSTHETIC GROUPS

In addition to the four levels of protein structure, some proteins can also have a **prosthetic group**. This is a non-protein group which is **strongly bound** to a polypeptide unit and it is essential for the protein's function. For example, myoglobin in muscle has a complex haem group which helps to draw oxygen into the muscle cells from the blood.

Myoglobin showing tertiary structure (blue) and the haem prosthetic group (grey) with oxygen attached (red).

THINGS TO DO AND THINK ABOUT

Most solids become liquid when they are warmed. So why does the egg-white protein, albumin, go solid when it is heated? The energy from the heat shakes the polypeptide chain and breaks all the weaker bonds of the secondary and tertiary structure so the hydrophobic R groups are exposed. The protein chains are now repelled by water and so clump together forming a solid mass.

Try adding strong acid to egg white – can you explain the effects? And what about whisking egg white – why does that change the protein structure?

CELL PROTEINS AND MEMBRANE STRUCTURE

IMPORTANCE OF THE R GROUPS

The R groups of amino acids are crucial to the structure and function of proteins.

1. R groups determine the **structure** of a protein. The primary structure means that the R groups have specific locations in the polypeptide chain. This determines the formation of regions of secondary structure in the chain, the interactions that can occur between R groups in the tertiary structure, the interactions with other polypeptide chains in the quaternary structure, and the interactions with prosthetic groups.

2. R groups can allow the **binding of ligands**. A ligand is a molecule that can bind to a protein. The R groups that are not involved in protein folding will be on the outer surface of the protein, so they are available for ligand binding. This is important because this binding changes the conformation of the protein which, in turn, changes its function or activity.

3. R groups at the surface of a protein determine its **location** within a cell. The balance of hydrophobic R groups and hydrophilic R groups (acidic, basic and polar) on the protein's surface will influence the solubility of the protein in the aqueous cytoplasm or interact with the hydrophobic layers of membranes (see below).

DON'T FORGET

The position of the R groups is determined by the primary structure.

CYTOPLASMIC PROTEINS

The cytoplasm is an aqueous medium so cytoplasmic proteins, such as enzymes and G-proteins, have to be soluble. To allow this, the surface of the protein has a greater proportion of hydrophilic R groups. The charges on the acidic and basic groups, and the slight charge of the polar groups can form weak interactions with the polar water molecules, making the protein soluble. During post-translational modification, other groups such as phosphates or sugars can be added to the surface of the protein to make it more hydrophilic. Hydrophobic R groups cluster at the centre of the folded polypeptide chain to form a globular structure with a hydrophobic core.

The charges on the surface R groups can be affected by the pH surrounding the protein; if there is no net charge on the surface so the protein is at its isoelectric point, it will be insoluble.

DON'T FORGET

Hydrophilic R groups predominate at the surface of a cytoplasmic protein.

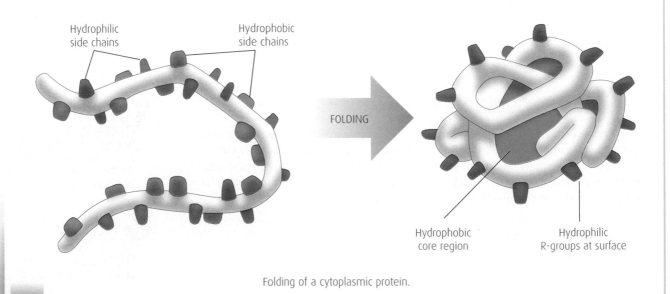

Folding of a cytoplasmic protein.

MEMBRANE STRUCTURE

Phospholipid bilayer

The membranes of cells are made of a **phospholipid bilayer** with globular proteins penetrating the bilayer or attached to its surface. This is called the **fluid-mosaic model of membrane structure** – the phospholipid bilayer can move so it is fluid, and the separate proteins are like the small pieces of a mosaic.

A phospholipid molecule has a hydrophilic head and hydrophobic tails. The phospholipids form a bilayer with the hydrophobic heads towards the aqueous cytoplasm and the aqueous external fluid, and the hydrophobic tails to the inside of the bilayer.

The hydrophobic centre of the phospholipid bilayer allows oxygen and carbon dioxide to pass through directly because they are **small non-polar molecules**. Hydrophobic signalling molecules (e.g. thyroxine and steroids) are larger molecules and, since they too are non-polar, they can also pass directly through the bilayer to reach their receptors in the cytoplasm.

The hydrophobic centre of the membrane acts as a barrier to the passage of **charged ions** (e.g. H^+, Na^+, amino acids) and most **polar molecules** (e.g. water or glucose). These can only cross the membrane through proteins suspended in the phospholipid bilayer.

Membrane proteins

The proteins in the membrane can be arranged in one of two ways.

- **Integral proteins** penetrate the hydrophobic interior of the phospholipid bilayer and are folded so that they have regions of hydrophobic R groups. These form **strong hydrophobic interactions**, tethering the protein to the tails of the phospholipids in the membrane's interior. Some integral proteins only extend partly into the bilayer, while others are **transmembrane proteins** that span the width of the membrane. The transmembrane proteins have many important roles in cells and organisms and form the basis of much of this unit.

- **Peripheral proteins** are not embedded in the phospholipid bilayer. Instead they form **weak bonds to the surface** of the membrane, either with the heads of the phospholipids or, more usually, to the exposed parts of integral proteins. Some peripheral proteins on the inside of the membrane are attached to the cytoskeleton, which helps to give mechanical support and shape to the cells.

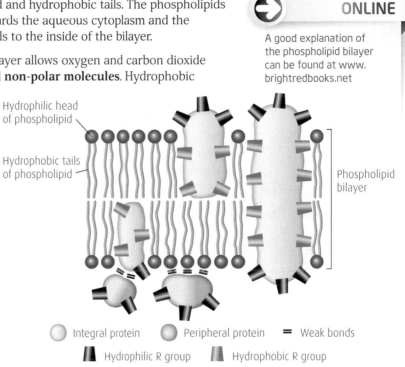

Hydrophilic head of phospholipid

Hydrophobic tails of phospholipid

Phospholipid bilayer

Integral protein   Peripheral protein   = Weak bonds

Hydrophilic R group   Hydrophobic R group

The fluid-mosaic model of membrane structure.

THINGS TO DO AND THINK ABOUT

Water moves across a membrane by travelling through a transmembrane protein called **aquaporin**. This has a pore through its middle, connecting the exterior liquid with the cytoplasm. Sketch out a diagram to show how the hydrophobic and hydrophilic R groups of the protein will be arranged to: 1) keep the aquaporin in the membrane, and 2) let water pass through the pore. You can check your ideas on page 28.

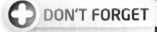

DON'T FORGET

Only non-polar molecules can pass through the hydrophobic region of the phospholipid bilayer.

ONLINE

A good explanation of the phospholipid bilayer can be found at www.brightredbooks.net

DON'T FORGET

Integral proteins interact with the hydrophobic interior of the membrane.

DON'T FORGET

Peripheral proteins form weak bonds to the surface of the membrane.

VIDEO LINK

Paul Anderson guides you through the structure of the membrane at www.brightredbooks.net

ONLINE TEST

Head to www.brightredbooks.net to test yourself on this topic.

BINDING AND CONFORMATIONAL CHANGE 1

Complementary shape. However, ligand-binding sites also have complementary chemistry and can change shape.

LIGANDS AND BINDING

Ligand is the general term for any substance that can bind to a protein. The binding of a ligand slightly changes the shape of the protein. How does this binding happen? And why is it important?

How ligand binding happens

The folding that produces the tertiary structure of a protein is stabilised by interactions between the R groups. However, there are many R groups that are not involved in the protein folding and these R groups may be exposed to the outer surface of the protein, facilitating binding to other molecules.

The protein folding produces **ligand-binding sites** on the surface of the protein. These binding sites are clefts in the surface of the globular protein which have a **complementary shape** to their ligand – they match the shape a bit like a jigsaw piece.

The binding site does more than match the shape of the ligand; it also has **complementary chemistry** to the ligand. The binding site has charged, polar and non-polar R groups arranged in it which match the charged, polar and non-polar areas on the ligand. The matched areas on the protein and the ligand interact, binding the ligand to the protein.

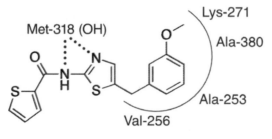

A ligand (black) binds to a protein. The numbers show the positions of the amino acids in the primary sequence of the protein. The ligand is interacting with the R group of Met-318 (hydrogen bonding) and with a hydrophobic non-polar pocket created by R groups of four other amino acids.

Why ligand binding is important

The interactions between the ligand and the binding site pull the polypeptide structure in towards the ligand, bringing about a change in shape – a **conformational change**. The shape of a protein is crucial to its function, so a change in conformation causes a **functional change** in the protein. Conformational change is the mechanism used to regulate the activity of proteins and is a recurring theme in the rest of this unit.

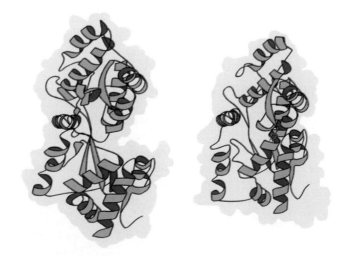

A ligand (red) binds to a protein, causing a conformational change and a functional change.

DNA AS A LIGAND

Some proteins have binding sites for parts of the DNA molecule. This means that DNA can bind to a number of different proteins and these have important roles.

1. **DNA packaging** in eukaryotic cells helps to form linear chromosomes. This packaging is achieved by wrapping the DNA in tight coils around **histone proteins** to form bead-like structures called **nucleosomes**.

   The DNA binds to the histone protein because the negatively-charged phosphates of the sugar–phosphate backbone interact with the positively-charged R groups arranged to the outside of the histone protein.

2. Other proteins control gene expression in a cell by **regulating the transcription** of genes. Each protein has a binding site which is **specific to a particular base sequence** on the double-stranded DNA. When one of these proteins is bound to the DNA, it can either stimulate the initiation of transcription (by facilitating the attachment of RNA polymerase) or inhibit the initiation of transcription (by blocking the attachment of RNA polymerase). Steroid hormones and thyroxine activate genes in cells by binding to these gene regulatory proteins.

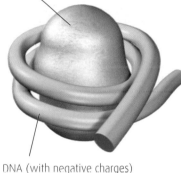

histone (with positive charges)

DNA (with negative charges)

A nucleosome.

VIDEO LINK

Watch how histones are used to package DNA into chromosomes at www.brightredbooks.net

ONLINE TEST

Test yourself on this topic at www.brightredbooks.net

THINGS TO DO AND THINK ABOUT

The nucleus contains DNA which is packaged with histone proteins. Until the 1940s it was thought that the genetic information would be carried by the more complex protein molecules as DNA was much too simple. Find out about the clever experiments of Avery and MacLeod.

BINDING AND CONFORMATIONAL CHANGE 2

ENZYMES AND LIGANDS

The active site of an enzyme is a ligand-binding site and the substrate is a ligand. When the substrate starts to bind to the active site, the small changes in the bonding pulls the enzyme's structure towards the substrate. This conformational change in the protein is called **induced fit** and it helps to further **increase the binding** and interaction between the active site and the substrate.

ONLINE

Watch an interactive tutorial on induced fit at www.brightredbooks.net

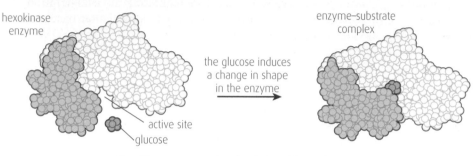

Hexokinase enzyme showing induced fit to glucose.

The new conformation is temporary and the enzyme it is trying to revert to the original conformation. This makes the catalysed reaction more likely to happen as the substrate is under tension – the **activation energy** of reaction is lowered by the stressing of the bonds in the substrate. Once catalysis takes place, the active site has less affinity for the products, which are released, and the enzyme returns to its original conformation.

Enzyme modulators

The rate of product formation by a metabolic pathway can be regulated by raising or lowering the activity of just one enzyme in the pathway. These enzymes are called **allosteric enzymes** (Greek for 'other shape') because their activity is regulated by altering their conformation.

This change in conformation is caused by a **modulator** that binds to the enzyme at a **secondary binding site** (also called an **allosteric site**). The change in enzyme shape alters the **affinity of the active site** for its substrate, thereby moderating the effectiveness of substrate binding and enzyme activity. Positive modulators increase the affinity and so increase enzyme activity, whereas negative modulators decrease the affinity and the activity.

DON'T FORGET

Modulators bind to an allosteric site away from the active site.

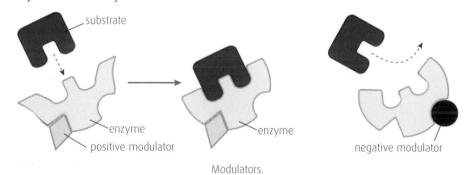

Modulators.

COOPERATIVITY

Some proteins with quaternary structure show **cooperativity** between their polypeptide subunits. **Ligand binding at one subunit** changes the conformation of that subunit; this new conformation alters the conformation of the neighbouring subunits, increasing the

contd

ligand **affinity of these remaining subunits**. In this way, an enzyme molecule made of a number of subunits (such as catalase) shows an increase in activity in the presence of its substrate – the binding of a substrate to the active site of one subunit causes conformation change in the other subunits so their active sites become active.

Cooperativity in haemoglobin

Without cooperativity, the oxygen saturation of haemoglobin would increase in a straight line as the oxygen concentration in the tissue increased. However, the oxygen dissociation curve is S-shaped, showing that the haemoglobin holds less oxygen in low oxygen surroundings and holds more oxygen in high oxygen surroundings. How can this happen?

Haemoglobin has four polypeptide subunits, each with an oxygen binding site. When an oxygen molecule binds to one subunit, it changes the conformation of the subunit. This changes the conformation of the other subunits, increasing the oxygen affinity of these other subunits. Because the oxygen affinity of a haemoglobin molecule increases further as oxygen binds to each subunit, oxygen collection is maximised where oxygen levels are high (in the lungs or gills).

Conversely, the release of an oxygen molecule from one subunit decreases the oxygen affinity of the other subunits. In low oxygen areas (in working tissues) the release of oxygen from each subunit decreases the oxygen affinity of the other subunits, thus maximising the release of oxygen where it is needed.

Hard working tissues generate heat. They also produce carbon dioxide, which reacts with water to form carbonic acid. The **higher temperature and lower pH** cause the haemoglobin to have a **lower affinity for oxygen**, so more oxygen is released into the tissues.

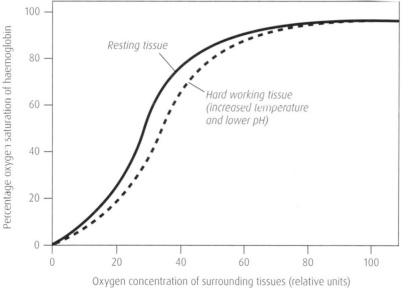

Oxygen dissociation curves. Compare the two lines at an oxygen concentration of 40 relative units – more oxygen is released from the haemoglobin to the hard working tissue.

DON'T FORGET

Cooperativity is when the binding of a ligand to one subunit of the protein increases the affinity of another subunit.

VIDEO LINK

Get to grips with haemoglobin cooperativity by watching the clip at www.brightredbooks.net

THINGS TO DO AND THINK ABOUT

Catalase enzyme is one of the most active enzymes and it shows cooperativity, so that it can rapidly remove the harmful hydrogen peroxide which has been made as a by-product of nucleotide synthesis. When hydrogen peroxide binds to one subunit of catalase, the other three subunits become active; there are now four active sites available to quickly break down hydrogen peroxide to water and oxygen.

ONLINE TEST

Test your knowledge on binding and conformational change at www. brightredbooks.net

REVERSIBLE BINDING OF PHOSPHATE

ALTERING PROTEIN ACTIVITY

A large subset of proteins is inactive when they are first synthesised and require **post-translational modification** to become activated. The most common form of activation is the **addition of a phosphate** on to the –OH of particular R groups (on serine, threonine or tyrosine). This phosphate can later be removed from the protein.

The addition or removal of a phosphate is used to cause **reversible conformational changes** in proteins and is an important method of **regulating the activity** of many cellular proteins, such as enzymes and receptor molecules. This ability to control protein activity is critical for signal transduction.

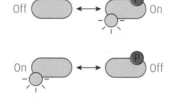

Phosphorylation can activate some proteins and deactivate others.

Kinases and phosphatases

The addition and removal of a phosphate from a protein is carried out by enzymes.

1. Kinases catalyse **phosphorylation**, which is the transfer of a phosphate from ATP to the protein. There are over 500 protein kinases in the human proteome.

2. Phosphatases catalyse **dephosphorylation**, which is the removal of a phosphate from a molecule. There are about 150 protein phosphatases in the human proteome.

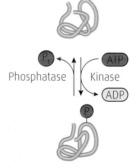

Phosphatase and kinase action.

Most protein kinases are dephosphorylated and inactive. They are activated by phosphorylation – by another kinase! This reversibility of kinase phosphorylation is important in **signal transduction cascades**, where kinases activate other downstream kinases to amplify a signal. Kinases and phosphatases are also crucial in the control of the cell cycle.

ATPases

ATPases are a group of transmembrane enzymes that use the phosphate from ATP to **phosphorylate themselves**, rather than their substrate. The phosphate binding changes their conformation and so alters their function. All transmembrane ATPases are involved in active transport of ions across the membrane, and the sodium–potassium pump (Na^+/K^+ ATPase) is an important example of this type of enzyme.

DON'T FORGET

The addition or removal of a phosphate can **activate** or **inactivate** a protein depending on how the conformation of the protein is changed.

DON'T FORGET

Kinases add phosphate on (KAPO), phosphatases encourage phosphate away (PEPA).

ONLINE

Hyperactive protein kinases are involved in cancers so kinase inhibitors are a major field of research. Find out about the many potential drugs that are in development at www.brightredbooks.net

MYOSIN AND ACTIN

These proteins are involved in muscle contraction and also in some movements that happen within cells, such as cytokinesis. The myosin protein has 'heads' that reach out and act as cross bridges when they bind to actin. The myosin moves the finer actin filaments in a series of 'power strokes' caused by conformational changes.

DON'T FORGET

Myosin does the moving, actin is acted on.

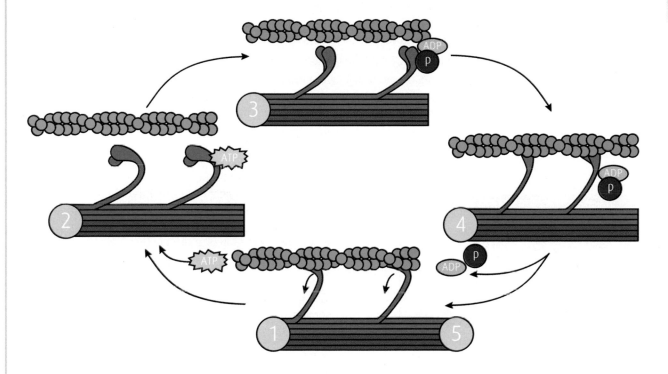

1. Myosin **heads** are bound to **actin** to form **cross bridges**

2. The **binding of ATP** to the myosin causes a conformation change. This makes the myosin head **detach from the actin** and **swing forwards**.

3. The ATP is broken down to ADP and phosphate. These stay bound to myosin and cause a second conformation change so …

4. … the **myosin head rebinds to the actin** which causes yet another conformational change in the myosin so …

5. … the **ADP and phosphate are released** from the myosin. The myosin head swings back to its first conformation, **dragging the actin filament** along. And then the cycle is ready to start again.

VIDEO LINK

The action of myosin is a dynamic process so it is easier to understand in a short clip at www.brightredbooks.net

THINGS TO DO AND THINK ABOUT

The ATP synthase enzyme found in mitochondria and in chloroplasts usually harnesses a proton gradient to regenerate ATP. However, like most enzymes, ATP synthase can work in reverse. This means that it can act as an ATPase, using the binding of a phosphate from ATP to pump protons against their concentration gradient.

ONLINE TEST

Head to www.brightredbooks.net to test yourself on the reversible binding of phosphate.

ROLES OF TRANSMEMBRANE PROTEINS

MOVEMENT OF IONS AND MOLECULES ACROSS MEMBRANES

DON'T FORGET

The movement of ions or molecules through channel proteins does not change the conformation of the protein.

Any ion or polar molecule that can pass across the membrane can only do so through a specific channel protein or transporter protein. For example, there are **specific transmembrane proteins** that act as sodium channels, glucose transporters or proton pumps.

The **control of ion concentration** on either side of a membrane is a key function of channel proteins and transporter proteins. The activity of some of these proteins can be altered so that they let more or less of their specific ion across the membrane, so changing the **concentration gradient** across the membrane.

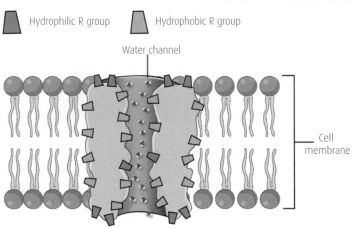

Hydrophilic R group   Hydrophobic R group

Water channel

Cell membrane

Aquaporin in cross-section, showing the key hydrophilic R groups in the pore that let water pass through.

Channel proteins

The passage of ions or molecules through channel proteins requires no change to the conformation of the protein channel. This is an example of **facilitated diffusion** because the proteins make it easier for the ion or molecule to move passively across the membrane. Each type of channel protein is **specific** for the ion or molecule that it allows through.

A good example of this is **aquaporin** which forms a selective channel to let polar water molecules pass through the pore. The surface of the protein has hydrophobic R groups, but some hydrophilic R groups help water molecules to pass through the narrow point at the centre of the pore.

Some channel proteins can be opened to allow diffusion of their ion, or closed to prevent diffusion of their ion – they are described as **gated channels** and there are two types.

Ligand-gated channels	Voltage-gated channels
Signal molecules bind to the ligand-gated channel protein and change its conformation.	If there is a large enough change in the ion concentrations across the membrane, this can change the conformation of a voltage-gated channel protein.
For example, to pass a signal on at a synapse, a neurotransmitter binds to a ligand-gated channel protein to trigger it to **open** and let Na⁺ ions flow through.	
In rods and cones in the retina, a ligand keeps a sodium channel **closed** until light activates a pathway to release the ligand and the channels lets ions through to generate a signal.	In the transmission of a nerve impulse along a nerve, the movement of ions across the membrane causes the voltage across the membrane to reach a critical level, so sodium channels **open**. The voltage change continues as more ions move across the membrane, which causes the channel to **close** again.

DON'T FORGET

Both types of gated channel open or close because the conformation of the protein is changed.

Transporter proteins

Like channel proteins, the transporter proteins are involved in **facilitated diffusion** and are **specific** for one type of ion or molecule. Unlike channel proteins, the transporter proteins bind to the ions or molecules, causing the protein to undergo a **conformational change** – the proteins actually pass the ions or molecules across the membrane, rather than just providing a route through. The GLUT4 glucose transporter is found in the membranes of fat and muscle cells and this transporter provides the route for facilitated diffusion of glucose into these cells.

DON'T FORGET

Transporter proteins undergo conformational change.

Another example is **coupled transport** whereby the movement of one material 'downhill' along its concentration gradient is used to pull another material 'uphill' against its concentration gradient. This is seen in the **glucose symport** process in the lining cells of the villi in the small intestine; sodium ions flow into the cell with their concentration gradient and glucose molecules are taken up against their gradient. The binding of the sodium ion and the glucose molecule to the symporter protein causes a conformational change that pushes the sodium ion and glucose across the membrane.

ONLINE TEST

Head to www: brightredbooks.net and test yourself on this topic.

contd

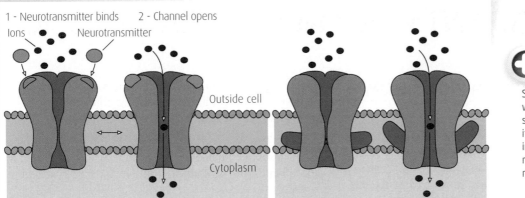

1 - Neurotransmitter binds   2 - Channel opens
Ions   Neurotransmitter

Outside cell

Cytoplasm

Ligand-gated channel being opened by the binding of a neurotransmitter ligand.

Voltage-gated channel being opened by a change in voltage across the membrane.

DON'T FORGET

Signal transduction occurs when an extracellular signal molecule binds to its receptor causing an intracellular response. The receptor can be in the membrane or inside the cell.

The lower concentration of sodium ions in the villus lining cell is maintained by the sodium–potassium pump (Na^+/K^+ ATPase) moving sodium ions out to the bloodstream. This Na^+/K^+ ATPase is an example of **active transport** because the conformational change in the protein requires energy from the hydrolysis of ATP. More accurately, the protein requires the binding of a phosphate from ATP.

Specialisation

To perform their specialised functions, different tissues in a multicellular organism are able to move different materials across their membranes. This specialisation in different cell types is due to having **different channel and transporter proteins** in their membranes. For example, glucose symporters are only found in the lining cells of the small intestine and the first part of the tubule of the nephron.

Similarly, the membranes of different cell compartments can have different channels and transporters. For example, the lysosome is a membrane bubble in a cell; its membrane has proton transporters which pump hydrogen ions into the bubble, forming an acidic fluid that breaks up materials inside.

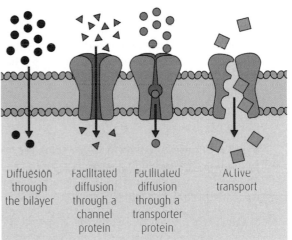

Diffusion through the bilayer | Facilitated diffusion through a channel protein | Facilitated diffusion through a transporter protein | Active transport

Diffusion is a passive process, requiring no ATP. Active transport needs ATP.

SIGNAL TRANSDUCTION

Some transmembrane proteins do not carry ions or molecules across the membrane. Instead, they are **receptor proteins** that have specific binding sites for **extracellular hydrophilic signal molecules**, such as peptide hormones and neurotransmitters.

The binding of the chemical signal to the receptor causes a conformational change in the protein. The part of the receptor protein that is in the cytoplasm changes shape and so it triggers a **signal-transduction pathway**, causing a specific intracellular response.

The signal-transduction pathway often involves cascades to amplify the signal by activating G-proteins or phosphorylation by kinase enzymes. The pathway can also change the uptake or secretion of molecules (as shown by insulin and ADH) or cause a rearrangement of the cytoskeleton.

Hydrophobic signal molecules (e.g. thyroxine and steroids) work by binding to receptor proteins **inside the cell** which then activate the transcription of specific genes.

ONLINE

Find out more about gated channels at www.brightredbooks.net

THINGS TO DO AND THINK ABOUT

Isotonic sports drinks contain a lot of salt and glucose. The high sodium ion content helps the glucose symport to pick up glucose from the intestine more readily, replenishing the lost glucose more quickly than just by drinking glucose solution alone. Now you know the biology, you can make an effective DIY sports drink for a fraction of the cost – search online for *Michael Mosley's sports drink*.

VIDEO LINK

A good animation to explain the different ways that materials can move across the membrane can be found at www.brightredbooks.net

ION GRADIENTS AND NERVE TRANSMISSION

THE MIGHTY SODIUM–POTASSIUM PUMP

The sodium–potassium pump (**Na⁺/K⁺ ATPase**) is a vital **transmembrane transporter protein** found in animal cells. The pump does **active transport**, moving sodium ions (Na⁺) out of cells and potassium ions (K⁺) into cells against their concentration gradients by using ATP to generate new ion gradients.

How it works

The sodium–potassium pump has **two stable conformational states**, one with a high affinity for sodium ions inside the cell, the other with a high affinity for potassium ions outside the cell. It will pump three sodium ions out of the cell (3 Na⁺ out) and two potassium ions into the cell (2 K⁺ in) to generate **ion concentration gradients** across the membrane.

Stage 1: In the first conformational state, the transporter protein has its ion binding sites exposed to the cytoplasm. The protein has a **high affinity for sodium ions** and **three Na⁺ ions bind** to the binding sites.

Stage 2: When the sodium ions are attached, the transporter protein is able to **hydrolyse ATP** to ADP and phosphate. The phosphate attaches to the protein to **phosphorylate** it and this causes a **conformational change** to the protein.

Stage 3: This new conformation has its ion binding sites exposed to the outside of the cell. Because it now has a **low affinity for the sodium ions**, Na⁺ ions are released outside the cell.

Stage 4: The new conformation has a **high affinity for potassium ions** and **two K⁺ ions bind** to the sites that are exposed to the outside of the cell. This triggers **dephosphorylation** – the release of the phosphate group from the protein.

Stage 5: This dephosphorylation causes the protein to revert to its **original conformation** with the binding sites exposed to the cytoplasm.

Stage 6: This conformation has a **low affinity for potassium ions** so K⁺ ions are released into the cell. The transporter protein has a high affinity for sodium ions again, so the cycle is back to Stage 1 and can repeat.

Why it's so important

The sodium–potassium pump is found in most animal cells and it accounts for about 25% of the basal metabolic rate of a human being!

DON'T FORGET

All animals rely on sodium–potassium pump pumping **2-K-in** (toucan).

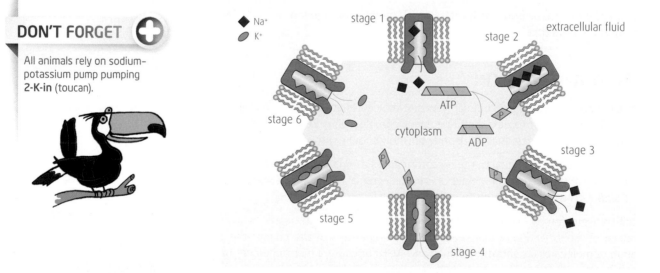

Six stages in the action of the sodium–potassium pump.

Why is the movement of sodium and potassium ions so important?			
Maintaining the **osmotic balance** in animal cells	Generation of sodium **ion gradient for glucose symport** in small intestine	Generation of sodium **ion gradient in kidney tubules**	Generation and maintenance of **ion gradients for resting potential in neurons**
Osmosis is affected by the solute concentration of the cytoplasm. The pump moves three Na⁺ ions out and two K⁺ ions into the cell, which lowers the overall ion concentration in the cytoplasm, and so increases the net water concentration.	The pump moves sodium ions out of the cells lining the small intestine to create a sodium ion gradient from intestine to lining cell. This enables the action of the glucose symport to absorb sodium ions and glucose from the intestine.	In cells lining the first part of the tubules, the pump moves sodium ions out of the cell to the blood. This creates a sodium ion gradient from the lumen of the tubule into the cell. This enables a glucose symport to absorb sodium ions and glucose from the filtrate.	See p.31.

NERVE TRANSMISSION

Resting potential of neurons

The pump moves three positively charged Na⁺ ions out of a neuron cell and two positively charged K⁺ ions into the neuron cell. The neuron membrane also has potassium channels which let some of the K⁺ ions leak back out of the cell, resulting in a net positive charge outside the cell. This imbalance in the electrical charge across the neuron membrane is called the **resting potential**. A nerve impulse passes along a neuron as a **wave of depolarisation** of the resting potential. How is this depolarisation triggered?

Triggering a nerve impulse

Depolarisation of the resting potential can be triggered by **neurotransmitter molecules** (e.g. acetyl choline) at a synapse. The neurotransmitter binds to a **transmembrane receptor protein** on the surface of the next neuron. This receptor is a **ligand-gated Na⁺ channel** so the binding of the neurotransmitter causes a change in conformation, making the channel open and letting Na⁺ ions diffuse through.

If **sufficient ion movement** occurs, then the voltage change across the membrane reaches a **critical level** and the membrane is **depolarised**. This voltage change causes the neighbouring **voltage-gated Na⁺ channel** to open, which leads to a domino effect as the triggering of one voltage-gated channel depolarises the membrane, so triggering the next voltage-gated channel and so on. And so the effect travels along the length of the nerve as a wave of depolarisation.

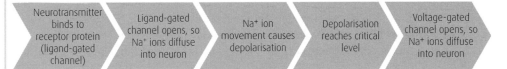

| Neurotransmitter binds to receptor protein (ligand-gated channel) | Ligand-gated channel opens, so Na⁺ ions diffuse into neuron | Na⁺ ion movement causes depolarisation | Depolarisation reaches critical level | Voltage-gated channel opens, so Na⁺ ions diffuse into neuron |

How a neurotransmitter triggers a nerve impulse.

Resetting the resting potential

After the wave of depolarisation passes along the neuron, there has to be a process to re-establish the resting potential ready for the next impulse. When the voltage reaches a critically high level, the **voltage-gated Na⁺ channel closes again** and **voltage-gated K⁺ channels open**. The K⁺ ions diffuse out of the neuron, in the opposite direction to the Na⁺ ions, restoring the resting potential. Once the resting potential is reached, the K⁺ channel closes again. Through all this, the sodium–potassium pump continues to work and the essential ion gradient is quickly reset.

| Voltage-gated channel opens, so Na⁺ ions diffuse into neuron | Voltage builds up so Na⁺ channel closes and K⁺ channel opens | K⁺ ions diffuse out of the neuron to reverse depolarisation | Resting potential is restored so K⁺ channels close again | Sodium-potassium pump resets the ion gradient |

How the resting potential is reset.

THINGS TO DO AND THINK ABOUT

Why does chilli taste 'hot' and mint taste 'cold'?

Heat thermoreceptors have a protein which undergoes a conformational change when temperatures reach 42°C or above. This opens an ion channel, which triggers a nerve impulse. Chillies have capsaicin which binds to this receptor molecule and cause the same conformational change!

Similarly, cold thermoreceptors have proteins which change conformation at 28°C or lower. Menthol binds to these receptors and this is what causes the 'cold' taste of mint!

Setting up the resting potential.
1) Na⁺/K⁺ ATPase moves ions.
2) K⁺ channels allow some leakage.

Na⁺
K⁺

DON'T FORGET

Nerve transmission is a wave of depolarisation due to Na⁺ influx into the neuron.

ONLINE

Find out why nicotine is addictive at www.brightredbooks.net

VIDEO LINK

The processes covered here are much easier to understand if you can see them happening as dynamic processes. Watch the sodium–potassium pump and the nerve impulse at www.brightredbooks.net

ONLINE TEST

Head to www.brightredbooks.net to test yourself on this topic.

DETECTING LIGHT

Light energy travels in the form of **photons**. Some organisms use the light energy from photons to generate ATP. Other organisms use the light energy to detect changes in their environment, so that the organism can respond and so increasing its survival chances.

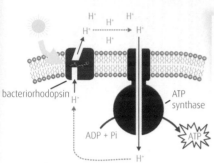

bacteriorhodopsin

ATP synthase

ADP + Pi

Bacteriorhodopsin generates a proton gradient which drives ATP synthase.

VIDEO LINK

Watch bacteriorhodopsin in action in the silent movie at www.brightredbooks.net

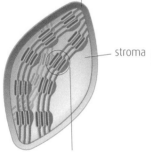

double membrane

stroma

stack of thylakoids (membrane pouches)

Chloroplast structure.

DON'T FORGET

Retinal combines with an opsin to form a photoreceptor protein.

PHOTONS AND ATP GENERATION

Two distinct methods of using light energy to generate ATP have evolved. Both methods use a proton gradient to drive ATP synthase and the difference is the way that the proton gradient is generated.

Archaea

Some archaea have **bacteriorhodopsin** molecules in their membranes. These transmembrane proteins have a prosthetic group called **retinal** which absorbs energy from photons. The energy absorbed by the retinal is used to **pump protons** through the bacteriorhodopsin, so that a **proton gradient** is generated across the membrane. The hydrogen ions then diffuse back through the membrane through **ATP synthase** to generate ATP from ADP and phosphate ions.

Plants

Plants (and other photoautotrophs) also generate a proton gradient, but the proton pumping mechanism is more complex. To keep the explanation simple, the energy from photons is absorbed by **photosynthetic pigments** in the chloroplast. The absorbed energy drives a **flow of electrons** along an electron transport chain and this electron flow is used to **pump protons across the thylakoid membrane** into the stroma. As with the archaea, the hydrogen ions then diffuse back through the membrane through **ATP synthase** to generate ATP.

DETECTING PHOTONS AND SIGNAL AMPLIFICATION

Retinal and opsins

The detection of light by all eukaryotes, from *Euglena* to elephants, uses the light-sensitive molecule called **retinal** to capture the light energy. Retinal is derived from Vitamin A, which itself can be formed from carotene in plants.

Retinal is a **prosthetic group** which is covalently bound to a polypeptide called **opsin**. This retinal–opsin complex is **embedded in membranes** inside **photoreceptor cells**. In vertebrates, the retina has two classes of photoreceptor cells, called rods and cones. The rod cells contain **rhodopsin** and are adapted to detect low levels of light, while cone cells have **photopsins** which allow colour vision.

Rhodopsin and the nerve impulse

Each rhodopsin molecule in a rod cell is connected to hundreds of **G-proteins**, which are the first part of a **cascade of proteins** that will amplify a photon signal and generate a nerve impulse. How does this happen?

The rod cell in darkness has a mechanism that prevents the generation of a nerve impulse. The rhodopsin is inactive and the rod cell produces **cyclic GMP (cGMP)**. This binds to ligand-gated Na^+ channels, thus opening the channels so sodium ions leak in across the membrane. The membrane stays depolarised and no nerve impulse is generated.

The rod could be thought of as having a cGMP 'handbrake' – activating the rhodopsin with light removes this handbrake so a nerve impulse can be generated (see p.33).

The protein cascade (rhodopsin/G-proteins/enzymes/channels) provides a **high degree**

contd

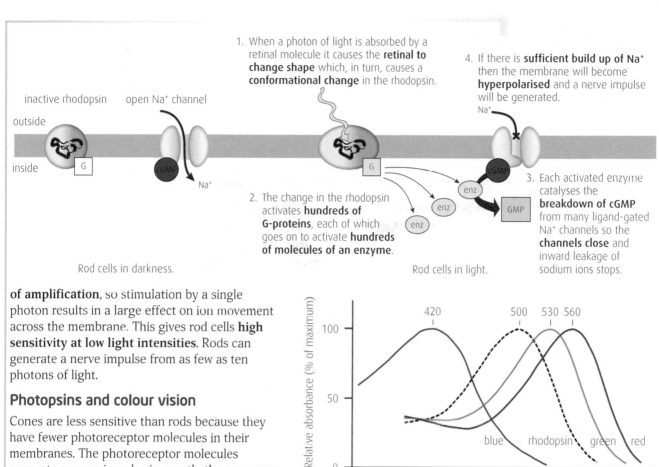

1. When a photon of light is absorbed by a retinal molecule it causes the **retinal to change shape** which, in turn, causes a **conformational change** in the rhodopsin.

4. If there is **sufficient build up of Na⁺** then the membrane will become **hyperpolarised** and a nerve impulse will be generated.

2. The change in the rhodopsin activates **hundreds of G-proteins**, each of which goes on to activate **hundreds of molecules of an enzyme**.

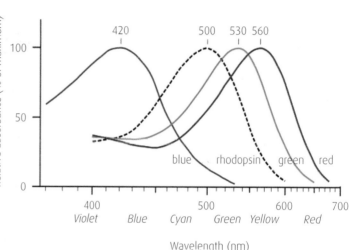

3. Each activated enzyme catalyses the **breakdown of cGMP** from many ligand-gated Na⁺ channels so the **channels close** and inward leakage of sodium ions stops.

Rod cells in darkness.

Rod cells in light.

of amplification, so stimulation by a single photon results in a large effect on ion movement across the membrane. This gives rod cells **high sensitivity at low light intensities**. Rods can generate a nerve impulse from as few as ten photons of light.

Photopsins and colour vision

Cones are less sensitive than rods because they have fewer photoreceptor molecules in their membranes. The photoreceptor molecules generate a nerve impulse in exactly the same way as in rods, but the cones have **photopsins** that are sensitive to different colours. In humans, each cone cell has one of three types of photopsin, each of which has a **maximal sensitivity to a specific wavelength** of light. In humans, one type of photopsin has a maximal sensitivity to blue, one to green and another to red light. Some species in other animal groups (such as insects, fish, reptiles and birds) have photopsins that have a maximal sensitivity to UV light.

Different photopsins are made by **combining retinal with different forms of opsin**. These opsin forms differ by a few amino acids in their primary structures, which means that their tertiary structures are different when they combine with the retinal. This change of structure means that photopsins absorb light across a different range of wavelengths.

Comparing the absorbance spectra of rhodopsin and human photopsins.

THINGS TO DO AND THINK ABOUT

The missing part of the **rhodopsin and the nerve impulse** story is the sodium–potassium pumps and permanently open K⁺ channels (see diagram). If you add these to the 'dark' diagram, can you see why the ion concentrations don't build up and the membrane remains depolarised? And if they are added to the 'light' diagram, can you see how the Na⁺ level builds up outside?

DON'T FORGET

Different photopsins in animals have maximal sensitivity to different wavelengths of light (red, green, blue, UV).

ONLINE

Read more about photoreceptor cells at www.brightredbooks.net

VIDEO LINK

Watch how activated rhodopsin generates the nerve impulse at www.brightredbooks.net

ONLINE TEST

Test your knowledge of detecting light at www.brightredbooks.net

COMMUNICATION WITHIN MULTICELLULAR ORGANISMS

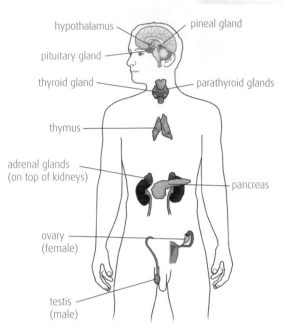

hypothalamus

pineal gland

pituitary gland

thyroid gland

parathyroid glands

thymus

adrenal glands
(on top of kidneys)

pancreas

ovary
(female)

testis
(male)

The main hormone-secreting endocrine glands in the human body. Each gland releases one or more specific signal molecule. Only certain tissues contain cells with the specific receptor proteins for these molecules. In this way, multiple signalling events can occur simultaneously. For example, the pancreas can release insulin, the pituitary gland can release anti-diuretic hormone and the thyroid can release thyroxine all at the same time, but the different target cells only respond to their intended message.

COORDINATION

In a multicellular organism, it is essential that cells communicate with one another. Without this communication the integration and coordination of cellular activities would be impossible. To achieve this communication, cells use **extracellular signalling molecules** and complementary **receptor proteins**.

The two principal forms of communication in multicellular organisms are **hormonal** and **nervous** (for the latter, see page 31). Both require the release of extracellular signals – that is one cell releases a signal molecule that the other cell detects and responds to. These cells may be close neighbours or may be at quite different locations within the organism. Either way, the signal has to leave one cell and travel to the other – the **target cell**. As these signals originate from outside the target cell, they are described as **extracellular**.

The same series of events occurs whatever the details of the signal, the target or the intended response. First, the target cells must have **receptor molecules** with a **binding site** for the **specific signal molecule**. When binding occurs, it **changes the conformation of the receptor**. This change in conformation **changes the behaviour of the target cell** in some way and this is called the **response**.

Different cell types produce **specific signals** that can only be detected and responded to by cells with the **specific receptor**. In a multicellular organism, different cell types may show a **tissue-specific response** to the same signal.

Hormones

Hormones are extracellular signalling molecules that are secreted by one tissue (such as an endocrine gland) into the blood. The hormone circulates in the bloodstream until it reaches the receptor protein of its target cell or until it is broken down. Hormones are either **hydrophobic**, such as the steroid hormones and thyroxine, or are **hydrophilic**, such as the peptide hormones ADH and insulin.

Neurotransmitters

The signals that are released into the synaptic gap between a nerve cell and its neighbour are called **neurotransmitters**. Nervous communication is very specific and rapid due to the intimate association between the signalling cell and its target cell.

Location of receptor proteins

Hydrophobic signalling molecules can **pass through membranes,** so their receptor molecules are **within the cytoplasm or nucleus** of the target cell. Hydrophilic signalling molecules cannot pass through membranes so require integral **cell-surface receptor proteins**.

DON'T FORGET

A hydrophobic signalling molecule is able to pass through membranes. A hydrophilic signalling molecule cannot.

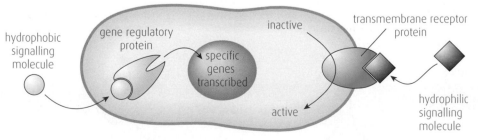

hydrophobic
signalling
molecule

gene regulatory
protein

inactive

transmembrane receptor
protein

specific
genes
transcribed

active

hydrophilic
signalling
molecule

A summary of the two signal transduction pathways.

HYDROPHOBIC SIGNALS

Hydrophobic signals and control of transcription

Hydrophobic signalling molecules include the **thyroid hormone thyroxine** and **steroid hormones**. Hydrophobic signalling molecules are lipid soluble, so they are able to move by diffusion across the hydrophobic part of the plasma membrane. The **receptors** for these signals are, therefore, deep in the cytoplasm or **within the nucleus** of the target cell. For example, the receptor proteins for steroid hormones are transcription factors. Only once the hormone signal has bound to the receptor can the transcription factor bind to gene regulatory sequences of DNA for transcription to occur.

Target	Tissue-specific response to thyroxine
Brain	Cortical development
Liver	Lower cholesterol
Pituitary	Inhibition of secretion of thyroid stimulating hormone
Multiple targets	Increase in metabolic rate

VIDEO LINK

Watch about steroid hormone signal transduction at www. brightredbooks.net

As a result, hydrophobic signals can **directly influence transcription** of genes.

Thyroxine

Thyroxine is a hormone that is released by the thyroid. It has a role in the control of the body's metabolism. There are many different tissue-specific responses to the release of thyroxine.

When thyroxine reaches a target cell, it is converted to thyroid hormone and stimulates transcription of certain genes, such as those for **Na⁺/K⁺ ATPase**. In the absence of thyroid hormone, the **thyroid hormone receptor protein** binds to the DNA and an associated co-repressor protein **inhibits the transcription of the gene**. When thyroid hormone binds to the receptor protein, conformational change prevents the co-repressor protein binding to the DNA and transcription of the gene for Na^+/K^+ ATPase occurs. With increased expression of Na^+/K^+ ATPase, the metabolic rate increases.

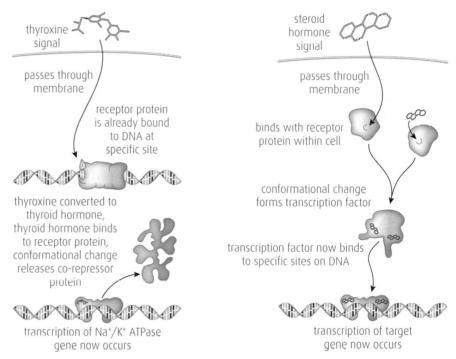

thyroxine signal

passes through membrane

receptor protein is already bound to DNA at specific site

thyroxine converted to thyroid hormone, thyroid hormone binds to receptor protein, conformational change releases co-repressor protein

transcription of Na⁺/K⁺ ATPase gene now occurs

steroid hormone signal

passes through membrane

binds with receptor protein within cell

conformational change forms transcription factor

transcription factor now binds to specific sites on DNA

transcription of target gene now occurs

Mode of action of thyroxine and steroid hormone. Both signals are hydrophobic so can pass through membrane to interact with receptor proteins within the cell. Once the signal molecules bind with their receptor proteins, the proteins change conformation. In the case of the steroid hormone, the receptor protein is a transcription factor that can now bind with the DNA and cause transcription to occur. In the case of thyroxine, its receptor protein is already bound to the DNA but the conformational change causes a co-repressor protein to be removed and this allows transcription to begin.

When transcription factors bind, they cause DNA to have lower affinity for histone proteins, so nucleosomes unwind allowing transcription to occur.

THINGS TO DO AND THINK ABOUT

Why would an increase in expression of Na⁺/K⁺ ATPase increase metabolic rate?

ONLINE TEST

Test yourself on this topic at www.brightredbooks.net

HYDROPHILIC SIGNALS

SIGNAL TRANSDUCTION

Hydrophilic signalling molecules, such as **peptide hormones** and **neurotransmitters**, are not lipid soluble. Hydrophilic signals cannot cross the hydrophobic part of the plasma membrane, they must instead bind to **cell-surface receptor molecules**. These **transmembrane proteins** change conformation when the signal ligand binds; **the signal molecule does not enter the cell** but the **signal is transduced** across the membrane of the cell. This means that the behaviour of the cell changes in response to the external binding of the signal molecule to the receptor molecule.

Transduced hydrophilic signals often involve **cascades activated by G-proteins** or **phosphorylation by kinase enzymes**.

Peptide hormones are small hydrophilic proteins. Well known examples include **insulin**, glucagon, somatotrophin (growth hormone) and **anti-diuretic hormone (ADH)**. Each one requires a specific receptor at its target cell surface. Since only target cells have the appropriate receptors at their surface, the action of these hormones is highly specific.

Examples of **neurotransmitters** include acetylcholine (ACh) and noradrenaline. Both are hydrophilic peptides. ACh is the transmitter at the neuromuscular junction connecting motor nerves to muscles. Noradrenaline has a role in the central nervous system and the sympathetic nervous system.

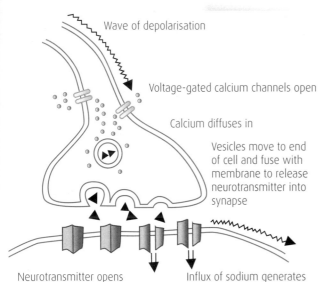

Wave of depolarisation

Voltage-gated calcium channels open

Calcium diffuses in

Vesicles move to end of cell and fuse with membrane to release neurotransmitter into synapse

Neurotransmitter opens ligand-gated sodium channels

Influx of sodium generates wave of depolarisation

A nerve impulse cannot cross a synapse. A signal molecule can, and this transduces the signal as a new nerve impulse.

NEUROTRANSMITTERS AND THE SYNAPSE

Neurotransmitters are the means by which nerves communicate across the synapse. The wave of depolarisation cannot spread directly from the membrane of one nerve to the next. Instead, the wave of depolarisation in one cell triggers the release of the **hydrophilic** neurotransmitters at the synapse and these signal molecules trigger the generation of a wave of depolarisation in the receiving cell.

When the wave of depolarisation reaches the synapse, **voltage-gated ion channels** open to allow calcium into the cell. This influx encourages the vesicles that the neurotransmitter is stored inside to move along the cytoskeleton and to fuse with the cell membrane. This releases the neurotransmitter, which diffuses across the gap. **Ligand-gated ion channels** are present in the cell membrane of the target cell. When the neurotransmitter binds, the channels change their conformation and this allows sodium ions to enter the cell. If sufficient sodium ions enter, then nearby voltage-gated channels generate a new wave of depolarisation.

INSULIN AND THE RECRUITMENT OF GLUT4

Blood glucose level must be kept within narrow limits, between 3.9 and 6.1 mmol per litre of blood. The **peptide** hormones **insulin** and glucagon interact in a negative feedback system to control blood glucose level. If the blood glucose level rises, for example after a meal, the pancreas detects the increase **in blood glucose**. This causes it to increase the secretion of **insulin**.

The **insulin receptor** is a **kinase-linked receptor** found in the cell membrane of fat and muscle cells. Once the signal binds, the signal is transduced and a series of **phosphorylation events** trigger the **recruitment of GLUT4 glucose transporters** to the cell membrane. These transporters allow glucose to enter these cells for further metabolism and the blood glucose concentration is controlled.

contd

DON'T FORGET ✚

There are three major types of hydrophilic signal receptor: ion-channel linked, kinase-linked and G-protein linked.

Diabetes mellitus is a medical condition caused by a deficiency in the effect of insulin and results in the loss of control of blood glucose level. It can be caused by **failure to produce insulin (Type 1)** or **loss of receptor function (Type 2)**. Type 1 diabetes is treated with injections of insulin. However, in Type 2 diabetes cells are no longer sensitive to insulin and this disease is generally associated with obesity. **Exercise** reduces the impact of Type 2 diabetes as it **triggers recruitment of GLUT4** through other metabolic pathways; **exercise improves the uptake of glucose** to fat and muscle cells in subjects with Type 2 disease.

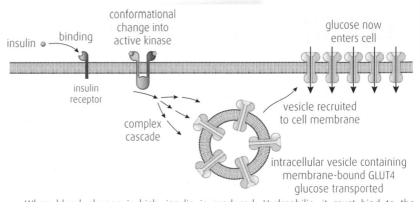

When blood glucose is high, insulin is produced. Hydrophilic, it must bind to the receptor at cell surface. The signal is transduced and a cascade of kinase-catalysed phosphorylation results in the recruitment of a vesicle to the cell surface. This vesicle is studded with GLUT4 glucose-transporter proteins. As soon as the vesicle becomes part of the cell membrane, glucose can enter the cell.

ADH AND THE RECRUITMENT OF AQP2

The **peptide hormone ADH** is released by the pituitary gland in response to low blood water concentration. The receptor protein is a **G-protein** coupled receptor in the cell membrane of the cells of the final part of the **kidney nephron tubule**, the collecting ducts. Binding of each ADH to its receptor in the collecting ducts of the kidney triggers the activation of many G-proteins. Each activated G-protein can, in turn, trigger multiple cascades of reactions. This amplification of the signal allows a rapid, localised and specific response. The ultimate response is the **recruitment of the channel protein aquaporin 2 (AQP2)** to the cell membrane that is in contact with the urine forming in the collecting duct. Aquaporins provide a highly efficient route for water to move across membranes and recruitment of AQP2 to this specific location allows **control of water balance in terrestrial vertebrates**. Failure to produce ADH or insensitivity of its receptor results in **diabetes insipidus,** a disease characterised by excessive watery urine.

There are several types of aquaporin along the nephron tubule. These recover water from the proximal convoluted tubule, the loop of Henlé, the distal convoluted tubule and the collecting ducts back into the blood capillaries (which have been moved away from the tubule in this diagram). The hormone ADH controls the presence of AQP2 in the collecting duct, and it is the presence or absence of this final aquaporin along the tubule that gives control of water balance.

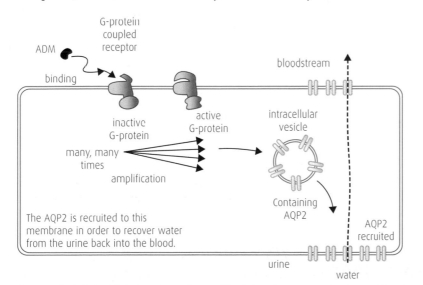

In the recruitment of AQP2 by ADH via a G-protein amplification system ensures a very rapid response – the control of water balance is critical to many cellular functions throughout the organism.

THINGS TO DO AND THINK ABOUT

Since the 1923 Nobel Prize for medicine was awarded to Banting and Macleod, many Nobel prizes have been awarded for research into aspects of cell signalling. To get a flavour of some of the challenges, rivalries and successes involved in the discovery of insulin, watch the feature film *Glory Enough for All*.

VIDEO LINK

Watch the animations about insulin, insulin receptors and diabetes at www.brightredbooks.net

REMODELLING THE CYTOSKELETON

It is possible to view organelles being moved by motor proteins travelling along the cytoskeleton in the laboratory. If a thin pond-plant leaf, such as that of *Elodea*, is viewed under bright field illumination, the chloroplasts are seen to move position. They are being moved by myosin proteins along actin filaments of the cytoskeleton. These are the same proteins that are involved in the contraction of animal muscle tissue.

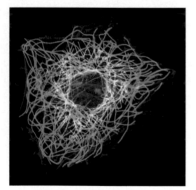

The three components of the cytoskeleton visualised under fluorescence microscopy. The microtubules are green, the intermediate filaments red and the actin filaments blue. *University of Basel*, imaged by R. Suetterlin, courtesy of C.A. Schoenenberger.

STRUCTURE AND FUNCTIONS OF THE CYTOSKELETON

The cytoskeleton is a **network of protein fibres** that extends throughout the cytoplasm in all eukaryotic cells. The cytoskeleton is attached to membrane proteins (see page 21) and gives **mechanical support and shape** to the cell, acting as scaffolding to maintain its shape. Organelles are attached to the cytoskeleton and it is involved in the **movement** of cellular components, such as vesicles and chloroplasts. The cytoskeleton is also responsible for the movement of whole cells – pseudopodia, flagella and cilia all rely on cytoskeletal activity.

There are three main components of the cytoskeleton: microtubules, intermediate filaments and actin filaments.

The progressive stages of crenation in a human red blood cell exposed to a hypertonic solution reveal evidence of the internal scaffold of the cytoskeleton.

Actin microfilaments

Actin microfilaments are **polymers** of the soluble **globular protein actin**. They consist of two strands of monomers wrapped around one another and have a thickness of only 2 nm. Actin microfilaments are responsible for **cellular movements**, such as contraction and pinching during cytokinesis in animal cells (see page 41), as well as the movement of cellular components.

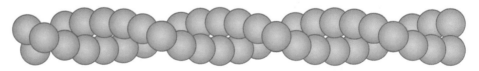

An actin microfilament is a two-stranded polymer formed from actin monomers.

Intermediate filaments

Rather than globular-protein monomers, intermediate filaments are formed from **fibrous proteins**. Rope-like in structure and about 10 nm in diameter, intermediate filaments consist of two pairs of monomers wrapped around one another and have tremendous **mechanical strength**. Only certain cell types contain intermediate filaments, such as the **keratin**-containing epithelial cells.

An intermediate filament is a tetramer of fibrous protein.

Microtubules

Microtubules are **hollow straight rods**. They are **polymers** of a dimer made from the soluble globular proteins **α-tubulin** and **β-tubulin**. The dimers are arranged to form microtubules and these have a diameter of 25 nm. The length of the microtubules is

contd

under the control of the cell through the addition (assembly) or removal (disassembly) of tubulin at the ends. Microtubules govern the **location and movement** of membrane-bound organelles and other cell components.

Found in all eukaryotic cells, microtubules radiate from the **microtubule-organising centre** (MTOC), which is located near to the nucleus and contains the **centrosome**. In animal cells, the centrioles form part of the centrosome and are involved in the organisation of the spindle fibres during cell division.

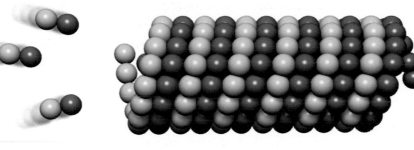

A microtubule is a polymer formed from 13 stacks of tubulin dimers.

THE MICROTUBULES AND MITOSIS

Cell division requires the remodelling of the cell's microtubules. The cytoplasmic microtubules are disassembled to form the spindle fibres and the spindle fibres control the **movement of chromosomes** during mitotic and meiotic divisions. The centrosome is duplicated and these then move to opposite ends of the cell. During cell division, the microtubules have three roles:

1. The **aster** or star-shaped tuft of microtubules at each centrosome ensures that the cell division apparatus is correctly located.

2. Some of the **spindle** microtubules consist of microtubules attached to kinetochore proteins at the centromeres of each chromatid.

3. Other microtubules in the spindle do not attach to the chromosomes but instead attach to microtubules from the opposite centrosome.

This arrangement allows the separation of the chromatids by the disassembly of the microtubules at the kinetochores, which has the effect of pulling the chromatids to the poles. The rate of assembly and disassembly of microtubules is **far higher** during cell division than at other stages in a cell's life.

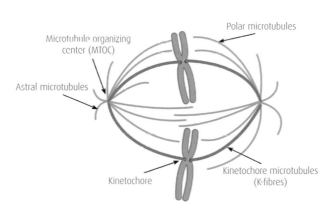

During mitosis or meiosis, the microtubules remodel themselves into the spindle. Some microtubules form the aster, which ensures the spindle is correctly aligned within the cell. Some attach to the kinetochore at the centromere and these spindle fibres are responsible for separating the chromatids. Other spindle fibres form the polar microtubules and these help to push the spindle apart during mitosis.

THINGS TO DO AND THINK ABOUT

Some drugs target microtubule action. Colchicine (from the autumn crocus) binds to soluble tubulin and prevents microtubule reassembly. It is used to induce polyploidy. Paclitaxel (from the yew tree) binds to tubulin in its polymer form and it prevents the disassembly of the microtubules. It is used in chemotherapy as it triggers cell death in cancerous cells with abnormally dynamic microtubule function.

ONLINE

Read about research into how loss of a dynamic cytoskeleton may be implicated in skin wrinkling and in Alzheimer's disease at www.brightredbooks.net

VIDEO LINK

Head to www.brightredbooks.net to watch a TED talk on this topic and an animation of the cytoskeleton.

DON'T FORGET

Microtubules and actin microfilaments are being constantly remodeled by adding monomers at one end and removing them from the other.

ONLINE TEST

Test yourself on this topic at www.brightredbooks.net

CELL CYCLE

THE SEQUENCE OF EVENTS IN THE CELL CYCLE

Some eukaryotic cells are able to reproduce themselves by following a sequence of events in which their contents are duplicated and divided in two. This **cell cycle** regulates the **growth** and **replacement** of **genetically identical cells** throughout the life of the organism.

The cell cycle consists of interphase and mitosis. The cell contents are built up during **interphase**, which has three phases called G1, S and G2. Interphase is followed by the M phase, which is composed of mitosis (the division of the nucleus) and cytokinesis (the division of the cytoplasm to form two cells). So the whole cell cycle is $G1 \rightarrow S \rightarrow G2 \rightarrow M$.

Interphase – G1, S and G2 phases

In actively dividing tissue, cells in interphase appear not to be doing anything when viewed by bright field microscopy. However, this is a very **active period** of growth and metabolism; interphase consists of an initial growth phase, G1, followed by an S phase, in which the cell continues to grow and copies its chromosomes, and a further G2 growth phase.

- **G1** is the first **growth stage** – the cell makes new proteins and copies of the organelles.

- **S** phase is when **DNA replication** occurs.

- **G2** is second period of **cell growth** – again, the cell makes more proteins and copies the organelles in preparation for mitosis.

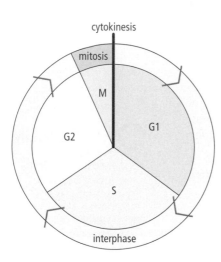

Cell cycle.

Mitosis and cytokinesis – the M phase

Mitosis is a dynamic continuum of sequential changes described as **prophase, metaphase, anaphase and telophase**. These phases flow from one to the next to divide the nucleus. **Spindle fibres** move the chromosomes during mitosis. These fibres are part of the cytoskeleton (see page 38).

- **Prophase** is the first visible sign of cell division. This is because the chromosomes **condense** (coil up) in preparation for being moved. Each chromosome is composed of two joined **chromatids** as DNA replication has already occurred. The cell's microtubules start to disassemble and start to assemble the **spindle fibres and asters** (see page 39). Some of the spindle fibres attach to the kinetochore proteins at the centromeres of the chromosomes. Tension-sensitive proteins at the kinetochore ensure that one of each chromatid pair is attached to each centrosome. The nuclear membrane disintegrates to facilitate the formation of two new nuclei.

- **Metaphase** involves the movement of the chromosomes to line up on the **metaphase plate at the equator** of the cell. This is an imaginary plane equidistant from the two centrosome poles. The movement is achieved by the assembly and disassembly of tubulin dimers into the microtubules forming the spindle fibres.

- **Anaphase** is a rapid phase that requires the disassembly of the kinetochore end of the spindle fibre. This pulls the **sister chromatids** apart. Once they are separated, the chromatids are called chromosomes in their own right.

- **Telophase** is the phase in which the cell briefly has two nuclei. The separated chromosomes are pulled by the spindle fibres to opposite poles to form **daughter nuclei**. The chromosomes start to uncoil and a nuclear membrane is made again.

- **Cytokinesis** is the division of the cytoplasm to form two **daughter cells**. Different mechanisms are used in animal calls and plant cells. In animal cells, the membrane is pinched in by a circle of actin and myosin fibres (see page 27; this process is known as cell cleavage). Plant cells have to form a middle lamella and new cell wall between the two daughter cells, and this is done before the membrane is made.

contd

Prophase: the chromosomes (blue in the photo) condense (coil up) and become visible as two joined chromatids. The cell's microtubules (green in the photo) start to form the spindle fibres. The spindle fibres attach to the kinetochores (pink in the photo) at the centromeres of the chromosomes. The nuclear membrane disintegrates.

Metaphase: spindle fibres move the chromosomes, so they line up on the **metaphase plate at the equator** of the cell.

Anaphase: the spindle fibres pull the **sister chromatids** apart. Once they are separated, the chromatids are called chromosomes in their own right.

Telophase: the separated chromosomes are pulled by the spindle fibres to opposite poles, to form **daughter nuclei**. The chromosomes start to uncoil and a nuclear membrane is made again.

Cytokinesis: the membrane is pulled in by actin and myosin (red in photo) to divide the cytoplasm to form two **daughter cells**. Plant cells have to form a middle lamella and cell wall before the membrane is made.

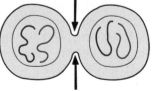

✚ DON'T FORGET

When you are asked about the cell cycle, it is important that you know the details of the interphase stages (G1, S and G2), as well as the stages of the M phase.

THINGS TO DO AND THINK ABOUT

The cell cycle can vary in duration depending on the tissue. A cell in the growing points of a mature plant may stay in G1 for weeks, whereas embryonic sea urchin cells can go through a complete cycle in two hours. When a sample of cells is viewed, the **mitotic index** can be found; this is the **percentage of cells undergoing mitosis** in the sample. If a tissue sample has an unusually high mitotic index, this may indicate a developing tumour.

✔ ONLINE TEST

Head to www. brightredbooks.net to test yourself on this topic.

CONTROL OF THE CELL CYCLE

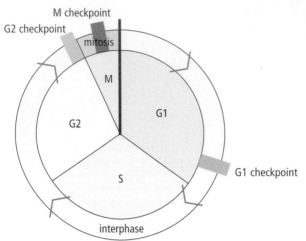

Cell cycle with checkpoints.

IMPORTANCE OF CELL-CYCLE CONTROL

The cell cycle has many complex events that all have to work perfectly to produce new daughter cells. Those new daughter cells have to be produced at the correct rate and in the correct locations to allow regulated growth and repair.

An **uncontrolled reduction** in the rate of the cell cycle may result in **degenerative disease**. In a degenerative disease, such as Alzheimer's or Parkinson's, insufficient replacement cells are being formed for normal tissue function.

An **uncontrolled increase** in the rate of the cell cycle may result in **tumour formation**. Uncontrolled growth may be benign or may result in a malignant cancer.

CELL-CYCLE CHECKPOINTS

Progression through the cell cycle is regulated by **checkpoints** at G1, G2 and metaphase. Checkpoints are critical control points where **stop** and **go-ahead** signals regulate the cycle.

The **G1 checkpoint** is near the end of G1. Here the **cell size is monitored**. There has to be sufficient cell mass to form two daughter cells. This checkpoint controls entry to the S phase.

If a go-ahead signal is not reached at the G1 checkpoint, the cell may switch to a **non-dividing state** called the **G0 phase**. In the G0 phase, cyclin proteins are not produced, so the cell has left the cell cycle and is said to be 'resting'. The cell will still be active and maybe a fully functioning differentiated cell, but it is no longer concerned with the business of cell division – most of your cells are currently in this 'resting' phase. G0 cells can return to the cell cycle if conditions change.

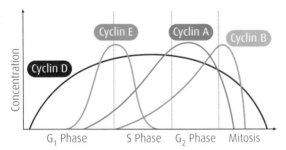

A series of different cyclin proteins are produced during the cell cycle. Each one is able to combine with a cyclin-dependent kinase (CDK) and, once its threshold has been reached, the cell cycle progresses to the next stage of the cell cycle.

The **G2 checkpoint** is at the end of G2. This assesses the **success of DNA replication** to make sure each daughter cell can receive a complete copy of the DNA. This checkpoint controls entry to mitosis.

The **M checkpoint** is during metaphase. This monitors the **chromosome alignment** to ensure each daughter cell receives one chromatid from each chromosome. This checkpoint controls the entry to anaphase. Thus, it triggers the **exit from mitosis and the start of cytokinesis.**

PROTEIN CONTROL OF THE G1 CHECKPOINT

Cyclin-dependent kinase

As the cell size increases during G1, **cyclin proteins** accumulate and combine with **regulatory proteins** called **cyclin-dependent kinases** (CDKs). The binding of the cyclin to the CDKs, forming what is known as the mitosis-promoting factor, results in the activation of the CDK. **Active CDKs** cause the **phosphorylation** of target proteins that stimulate the cell cycle. If a sufficient threshold of phosphorylation is reached, the checkpoint is passed with a 'go ahead' and the cell cycle moves on to the next stage. If an **insufficient threshold** is reached, the cell is held at a checkpoint, as 'stop' is the default setting (except in cancer cells). Without any active CDKs the cell will enter the G0 resting state.

DON'T FORGET

A kinase is an enzyme that phosphorylates a protein substrate.

contd

Retinoblastoma protein

The **retinoblastoma protein** (Rb) is an important part of the G1 checkpoint. It is a **transcription-factor inhibitor**. These transcription factors are involved in the production of proteins that are required for DNA replication in S phase. With low levels of CDK activity, Rb binds to the **transcription factor E2F**. This **binding of Rb to the transcription factor inhibits the transcription of the genes** required to enter the S phase. Inhibition of the transcription factors is the molecular stop signal of the checkpoint.

If the activity of CDK reaches a **sufficient threshold** where each Rb has been repeatedly phosphorylated (14 times), this **phosphorylated Rb** protein can no longer bind to the **transcription factors**, releasing them to promote the transcription of genes required for DNA replication in the S phase. This phosphorylation of Rb, by preventing its binding to and therefore inhibition of the transcription factors, allows the go-ahead signal to be given at the checkpoint.

Lack of regulation of this CDK–RB–E2F pathway has been detected in almost all malignant cancer cells in humans.

Cyclin level	Absent	Present	Abundant
CDK status	Inactive	Active but in low concentration	Active and in high concentration
Rb protein	Unphosphorylated	Rb with single phosphorylation	Rb with multiple phosphorylation
Transcription factors for DNA replication	Not produced in significant quantity	Transcription inhibited (as TFs present but bound to Rb)	Transcription promoted (as TFs present and not bound to Rb)
Cell-cycle stage	Cell enters G0 (differentiation may be triggered)	G1 checkpoint 'STOP'	G1 → S 'go ahead' at G1 checkpoint

p53 protein

The **p53 protein** is another important part of the G1 checkpoint. It is a **transcription factor** that can stimulate DNA repair, arrest the cell cycle or trigger cell death.

If DNA damage has occurred p53 causes the expression of genes that **stimulate DNA repair** and **arrest the cell cycle**. If the DNA repair is successful, the cell cycle can continue once again.

If the damaged DNA cannot be repaired, then p53 instructs the cell to kill itself through **apoptosis** (see page 44).

The p53 protein has been found to be missing or faulty in about 50% of human cancerous cells – one reason why these cells can divide without being halted at the G1 checkpoint.

THINGS TO DO AND THINK ABOUT

The genes that are responsible for abnormal cell division are known as proliferation genes and anti-proliferation genes.

Proliferation genes code for proteins that promote cell division, such as CDK genes. Under normal circumstances this would happen only when an external signal is received, for example when a wound needs to be repaired. They are also known as proto-oncogenes because, when they mutate, they form oncogenes, the genes that are found in many cancers. These mutations result in proteins that stimulate excessive cell division and tumour formation, even in the absence of any external signal. Because this mutation only needs to happen in one gene of the pair, oncogenes can be thought of as dominant genes.

Anti-proliferation genes code for proteins that restrict cell division by operating at the cell-cycle checkpoints. Examples would be Rb and p53. Their genes are also known as tumour-suppressor genes because they normally prevent excessive cell division. One functioning gene can still produce the protein to inhibit cell-cycle progression, so both copies of the gene must mutate before control of the cell cycle is lost and a tumour starts to form. For this reason, the cancer-causing mutations of anti-proliferation genes can be thought of as recessive.

DON'T FORGET

Remember the exact timings of the checkpoints. The G1 checkpoint is near the end of G1 phase. The G2 checkpoint is at the end of G2 phase. The M checkpoint is during metaphase.

ONLINE TEST

Test yourself on this topic at www.brightredbooks.net

APOPTOSIS

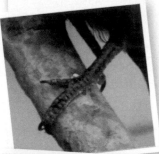

Most members of the bird order Coraciiformes have syndactylous feet. This means that two of their three forward-pointing toes on each foot are fused together, as can be seen in this grey-headed kingfisher from Tanzania. The cells between the two fused toes (the middle and the outer toe) have not undergone apoptosis during development (unlike the cells between the middle and the inner toe). Kingfishers are sit-and-wait predators and it is thought that the fused toes provide additional stability for long periods spent immobile.

PROGRAMMED CELL DEATH

In a multicellular organism, growth and development involves not only the careful control of cell division by the cell cycle, but also the careful control of the deliberate destruction of cells by **apoptosis**. For example, in the development of tetrapod limbs, different patterns of programmed cell death result in a different arrangement of fused or free fingers or toes. In mammalian cell cultures, primary cell lines undergo apoptosis after about 60 cell divisions; cancer cell lines that have mutations in their regulation of apoptosis have to be used instead to achieve immortality.

TRIGGERING APOPTOSIS

Programmed cell death (apoptosis) is triggered by **cell-death signals**. These can come from outside of the cell or from within the cell. It is important that the rate of apoptosis is carefully controlled. If the degradation enzymes called **caspases** that achieve apoptosis are overactive, then neurodegenerative diseases can result. On the other hand, underactive caspases can be a cause of tumour development.

The extrinsic pathway

Cell-death signals may originate outwith the cell, for example from a lymphocyte, such as a killer cell that has detected abnormal behaviour of the cell. These **ligands bind** to a surface-receptor protein, causing the receptor protein to **change the conformation** of its subunits on the cytoplasmic side of the membrane. This change acts as a **signal transduction** and activates a **protein cascade** that produces **caspases**.

The intrinsic pathway

Death signals may also originate within the cell, for example as a result of DNA damage the presence of **p53 protein** can also activate a **caspase cascade**. This is achieved through the disruption of the integrity of the mitochondria, which triggers the cascade. Cells may also initiate apoptosis in the absence of cell growth factors.

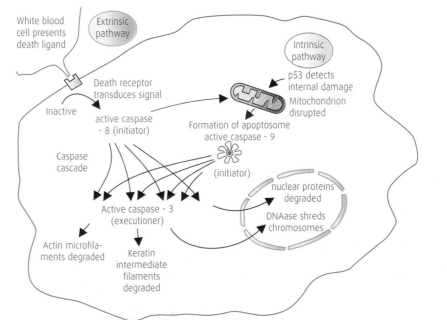

A simplified summary of the cellular pathways that lead to programmed cell death. Note that, whether the starting point was intrinsic or extrinsic, both pathways lead to the same end point. Once the caspase cascade has activated the executioner caspases, the cell's cytoskeleton and nuclear material is shredded.

THE CASPASE CASCADE

The caspase cascade involves a series of **post-translational modifications** to proteins that are already present in the cytoplasm. This allows a **rapid response** to occur, even in circumstances when the cell's nucleus is out of action. The post-translational modifications involve **cleavage** – subunits are removed and they can reassemble to form active caspases. The caspase cascade involves **inactive caspases** becoming **activated**. Each 'initiator' caspase at the beginning of the cascade activates several **executioner caspases**. These act as DNAase, proteinases and enzymes that destroy keratin and actin filaments. Each executioner caspase degrades over 600 cellular components and this growing cascade destroys the cell.

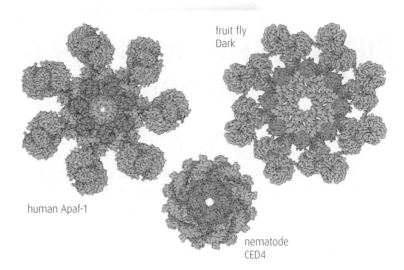

fruit fly
Dark

human Apaf-1

nematode
CED4

The human apoptosome, Apaf-1, is a molecule involved in the cascade that leads to a cell's death. Similar molecules with a similar function are found in other animals. It may be that specific differences in the proteins involved in apoptosis will allow us to develop more targeted drug treatments for tackling parasite infections and their diseases.

FRAGMENTATION OF THE CELL

Once the cellular components have been destroyed, the cell can no longer function. The lack of intact cytoskeleton means that the cell breaks into many small vesicle fragments. These are rapidly engulfed and digested by phagocytic white blood cells.

Time-lapse images showing the *in vitro* apoptosis of a prostate cancer cell treated with the chemotherapy drug *etoposide*. This drug causes DNA damage within the cells, inducing programmed cell death. The fragmentation of the cell is diagnostic of apoptosis and, in a whole organism, allows rapid engulfment by macrophages.

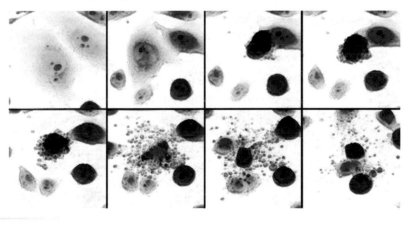

THINGS TO DO AND THINK ABOUT

The interaction of parasites with apoptosis is worthy of further research. Many parasites induce apoptosis in their host cells as part of their lifecycle. The fever symptoms of malaria are the result of the asexual amplification cycles of *Plasmodium* parasites simultaneously rupturing their host's red blood cells by inducing apoptosis. Other parasites may inhibit apoptosis in their host cell. For example, the endoparasite *Toxoplasma gondii* inhibits the host's intrinsic pathway, preventing apoptosis and providing a longer term niche for parasite survival and asexual reproduction.

However, the fact that parasite cells can also exhibit apoptosis may lead to potential new drug targets. While the apoptotic pathway is relatively well conserved between species, differences may provide the potential for the development of novel drugs; it may be that the problems of resistance to anti-helminth drugs will be sidestepped by drugs designed to induce apoptosis in *Schistosoma*. Even the protozoa, such as *Plasmodium*, have meta-caspases which are sufficiently different from our own caspases to provide good future drug targets.

ONLINE TEST

Revise this topic by taking the test at www.brightredbooks.net

FIELD TECHNIQUES FOR BIOLOGISTS: HEALTH AND SAFETY

Fieldwork is practical work carried out by a researcher outside of the laboratory or office. As you will have realised already, variables are much harder to control in a natural environment than in a laboratory; this adds additional complexities to the **identification of hazards** and **minimisation of risks** in a field setting. The *Health and Safety at Work Act* requires all hazards and risks to be removed or, where that is not possible, to be reduced as far as is reasonably practicable by control measures. This applies to fieldwork, as much as to any other type of work, and various legislation covers teachers, pupils, students and volunteers engaged in fieldwork.

HAZARDS ASSOCIATED WITH FIELDWORK

Fieldwork may involve a **wider range of hazards** compared to working in the laboratory. There are three considerations that are particularly relevant when assessing fieldwork for risks:

- terrain
- weather conditions
- isolation.

Terrain

Terrain refers to the type of ground where the fieldwork is conducted, and this will dictate the safest type of footwear, and whether harnesses or other safety gear is required. Slipping and tripping are the single most common cause of major injury in UK workplaces, so the hazards associated with uneven ground must be considered. Working outdoors on mountains, near cliffs, in trees, in mud, in water, or on ice pose particular hazards and to minimise their risks appropriate precautions would be taken. For example, without specialist training and equipment to reduce risk of injury, fieldwork in the canopy of tall trees or on cliffs is not possible.

Weather conditions

Even areas with normally innocuous terrain can become hazardous in severe weather. Any fieldwork in the hills of Scotland, for example, requires close consideration of the hazards associated with cold, wet or windy weather. The hazards associated with sun exposure or electrical storms also must be considered. The latter is a good example of a threat that is a low risk but a very high hazard.

Isolation

If the fieldwork site is a long distance from habitation, medical help or transport links, this can pose a hazard. Moving supplies or equipment will be more difficult and evacuation procedures will have to be considered in the event of injury or illness. Hazards and risks associated with lone working also need to be considered.

A small part of the risk assessment documentation for a fieldtrip.

RISK ASSESSMENT OF FIELDWORK

Risk assessments should have a level of detail that is appropriate to the types of hazards and the size of the risks. Risk assessments that are cluttered with minor concerns are not helpful as they are less likely to be followed in practice.

contd

ONLINE

To see an example of a Health and Safety policy for university fieldwork, head to www.brightredbooks.net

Even the best risk assessment cannot prevent all incidents. While on a field trip, one of the authors was stung on the head by this box jellyfish. This painful experience was successfully treated with vinegar, which had been included in the medical kit just in case this very unlikely event occurred!

A suitable and sufficient risk assessment

- identifies significant foreseeable hazards associated with the field trip, including travel to and from the location

- evaluates the associated risks (how likely)

- evaluates the severity of the hazards (how harmful)

- identifies appropriate control measures to reduce the level of risk or hazard

- records these safe working practices.

Conducting fieldwork at night requires careful risk assessment. If the risks can be reduced to an acceptable level, it is possible to observe species that are not active during the day, such as the (a) Oriental bay-owl and (b) Malayan colugo, both photographed on a field trip to Tioman in Malaysia. The owl was the first recorded on the island and this sighting added to our understanding of the distribution of this very rarely observed species.

ONLINE TEST

Test yourself on the topic of health and safety at www. brightredbooks.net

THINGS TO DO AND THINK ABOUT

Consider what different hazards and risks you would have to assess before conducting biological fieldwork in an urban city park, versus a small offshore island.

FIELD TECHNIQUES FOR BIOLOGISTS: SAMPLING OF WILD ORGANISMS

Sampling is the gathering of data from part of a population. Sampling can eliminate the need to measure and record every member of a population, as long as the sample selected is representative of the variation in the population being studied (see page 98).

In many studies, the sampling of wild organisms can be achieved through the use of **observations.** In other cases, the organisms have to be **captured** and they may or may not be **released** later, depending on the type of study. Whatever the study design, the sampling must be carried out in a manner that **minimises any impact** on wild species and habitats.

ONLINE

Before sampling wild animals in Scotland, ensure that you have checked the Scottish Natural Heritage website to see whether the species or area is protected. You can find the link at www.brightredbooks.net

The oysterplant, here on a beach in the Northern Isles, has a restricted niche and is rare in Scotland. While there are hundreds of species that are protected in Scotland, surprisingly, this is not one of them. This protected status does differ across the UK, however, so anyone sampling this species would need to be aware of the local differences in legislation.

MINIMISING THE IMPACT OF SAMPLING

In any scientific study, it is important that the measurement method selected does not influence the value of that measurement. In sampling, scientists must consider whether the sample being measured will be affected by the sampling procedure and, if it is, whether that will have any impact on the results being gathered and any potential conclusions being made. If sampling is invasive and involves the temporary or permanent removal of individuals from the population, the benefits of the study must outweigh any negative impact of sampling.

A common lizard in the Cairngorm National Park. The handling of lizards is not recommended, especially by people who are unable to distinguish between the species that have special legal protection and those that do not. In areas that have protected-status designation, such as a National Park, there may be increased protection of particular species or habitats.

In particular, it is essential to consider whether the sampling involves **rare species** or **vulnerable habitats** that are protected by legislation. Rare species are those that are found at low density or those whose total population is small. Vulnerable habitats are those that are easily damaged by human activities. Sampling in either of these cases may be illegal, unless carried out under license.

SAMPLING TECHNIQUES

The chosen technique must be appropriate to the species being sampled.

Point count

A point count is carried out from a selected stationary location and involves the gathering of observational data. It is used for determining species abundance and is a technique that is often used in sampling bird populations.

Transect

Transects are used for determining changes in community across an environmental gradient, such as a shore. Transects can be in the form of a **line**, which involves a narrow focus of sampling, or a much wider **band**. Whether line or band transects are more applicable depends on the variability of the communities. The

Point counts for endangered water birds in Cambodia. (a) The use of optical aids allows observations to be made without disturbing the birds. (b) Closer approach causes the birds, such as these Oriental darters, to flee and this reduces the accuracy of the sample data gathered.

contd

higher the variability at each point along the transect, the wider the band has to be in order to make the transect representative of the communities being sampled.

Remote detection

In this type of sampling, the monitoring is carried out at distance using sensors, for example by satellite. This is used for global vegetation surveys or for gathering data in areas that are difficult to access.

Students using different sampling strategies in Perthshire. (a) Line transect sampling of the plant communities involved in the succession of a pond. The students are using quadrats systematically every 5 m along the transect to estimate the abundance of each plant species. (b) Comparing the biodiversity of grasslands of different grazing intensities. The students are using random location quadrat sampling within defined areas of each field, which is good practice. However, the choice of quadrat type is poor – can you anticipate what difficulties were encountered through the use of a gridded quadrat?

Quadrats

The use of quadrats ensures that a standard area is being sampled each time a measurement is made. This helps to ensure an equal amount of effort is applied at each sample point, which helps to reduce bias and increase reliability. Quadrats are particularly suitable for sessile (fixed) and slow-moving organisms. The appropriate size of quadrat is determined by how uniformly organisms are distributed in the habitat. Smaller quadrats are suitable when the organisms are small or are very densely packed. The diversity influences the number of samples that will need to be taken – more samples are required with higher diversities. The sample shape is not important, as long as it is uniform and that the area of the quadrat is known.

Mobile species

The sampling of mobile species may involve capture techniques, such as nets or traps. Birds and bats can be caught in mist nets, but only under license. Night-flying insects can be sampled using a light trap. Small mammals can be caught in Longworth traps. All of these techniques allow the animal in question to be released unharmed.

Elusive species

Some animals are very difficult to sample through normal observation. Instead, camera trapping can be used to provide direct evidence, or scat sampling (counting droppings) can be used to provide indirect evidence.

A sample of blood being taken from a hermit thrush in a study of the metabolism of migrant birds. Mist nets were used to trap a sample of this mobile species. Once the birds were individually ringed, a subsample was selected for the study.

Some animals are elusive and difficult to sample. (a) This leopard in Tanzania was extremely difficult for observers to see, even when it was pointed out by others. (b) The large and obvious Leopard scats, on the other hand, are difficult to miss.

ONLINE TEST

Head to www. brightredbooks.net to test yourself on the sampling of wild organisms.

THINGS TO DO AND THINK ABOUT

Key legislation that you may need to be aware of before sampling wild organisms in Scotland includes the *Wildlife and Countryside Act 1981*, the *Habitat Regulations 1994* and the *Protection of Badgers Act 1992*.

FIELD TECHNIQUES FOR BIOLOGISTS: IDENTIFICATION

THE CLASSIFICATION OF LIFE

The study of biology is made particularly challenging by the vast diversity of species. As yet there is no real consensus figure for global biodiversity, but it is estimated that 1.5 million species have already been named and described. In addition, estimates suggest that there may be another 10 million species living on the planet. On top of that figure, are the billion species that are likely to have existed on the planet over the last 4.5 billion years that are now extinct.

Of course, these millions of species are all related and are all descended from the last universal ancestor that lived on Earth about 3.5 million years ago. In this time, mutation, evolution and speciation have resulted in the diversity we see now. To help us understand our observations of an organism (or of its cells, DNA or proteins) it is helpful to **identify** which one of the million living species we are considering. In this way, we are able to collate and compare information gathered about each species. By understanding **taxonomic relationships** between different species, we can make deductions and predictions about how applicable knowledge gained about one species is to another. In this way, the classification of life according to relatedness is central to biological understanding.

DON'T FORGET

It is possible to identify species without understanding taxonomy. It is also possible to have a good idea of the taxonomy of an organism that cannot be specifically identified.

A recent study of the classification of European warblers has led to the recognition of a previously unrecognised species, the Moltoni's warbler. This species breeds in the Mediterranean and spends the winter in Africa, but occasionally overshoots further north in spring. This individual was spotted on Fair Isle and was identified in the field using the tentative information in the latest classification guides. After being trapped for ringing, a biological key was used to confirm the identity. DNA analysis was also used, which supported the identification, and this has been used to help validate the characteristics in the classification guide.

IDENTIFICATION

Identification is the ability to put a species name to a sample. Identification of a sample can be made using classification guides or biological keys, or by analysis of DNA or protein.

contd

Classification guides tend to focus on one class or order of organisms from one geographical location, such as a guide to the birds of Europe. These guides provide information that allows separation of similar species, such as descriptions of key features, illustrations, habitat preferences, known distributions and seasonal abundance.

A classification guide for American warblers.

More specialist identification information can be found in the form of **biological keys**. These pose a series of questions, often in the form of paired alternative statements, which focus only on the characteristics that allow different species to be separated. In a paired-statement key, each option leads to another paired statement or to the conclusive identification of a species.

The **analysis of DNA or protein** is a powerful technique for the separation of species that are otherwise only subtly different to one another. In addition, it allows the identification of species when only partial or molecular evidence remains.

Section H: Pelvic Fins Thoracic With Less Than Five Soft-rays (continued)

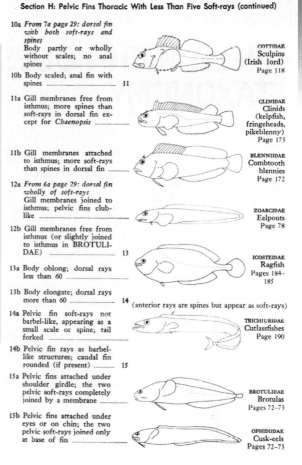

10a From 7a page 29: dorsal fin with both soft-rays and spines Body partly or wholly without scales; no anal spines ....
COTTIDAE Sculpins (Irish lord) Page 118

10b Body scaled; anal fin with spines 11

11a Gill membranes free from isthmus; more spines than soft-rays in dorsal fin except for *Chaenopsis*
CLINIDAE Clinids (kelpfish, fringeheads, pikeblenny) Page 173

11b Gill membranes attached to isthmus; more soft-rays than spines in dorsal fin ....
BLENNIIDAE Combtooth blennies Page 172

12a From 6a page 29: dorsal fin wholly of soft-rays Gill membranes joined to isthmus; pelvic fins club-like ....

12b Gill membranes free from isthmus (or slightly joined to isthmus in BROTULI-DAE) 13
ZOARCIDAE Eelpouts Page 78

13a Body oblong; dorsal rays less than 60 ....
ICOSTEIDAE Ragfish Pages 184-185

13b Body elongate; dorsal rays more than 60 14

14a Pelvic fin soft-rays not barbel-like, appearing as a small scale or spine; tail forked ....
(anterior rays are spines but appear as soft-rays) **TRICHIURIDAE** Cutlassfishes Page 190

14b Pelvic fin rays as barbel-like structures; caudal fin rounded (if present) 15

15a Pelvic fins attached under shoulder girdle; the two pelvic soft-rays completely joined by a membrane ....
BROTULIDAE Brotulas Pages 72–73

15b Pelvic fins attached under eyes or on chin; the two pelvic soft-rays joined only at base of fin ....
OPHIDIIDAE Cusk-eels Pages 72–73

A biological key for coastal marine fish.

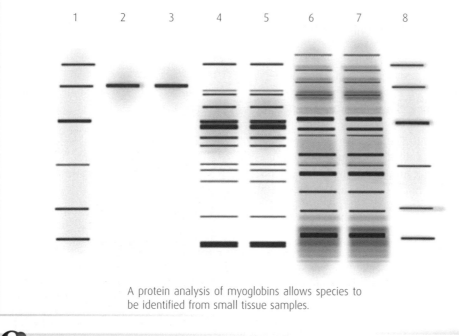

A protein analysis of myoglobins allows species to be identified from small tissue samples.

ONLINE

A link to the fabulous iSPOT can be found at www.brightredbooks.net

THINGS TO DO AND THINK ABOUT

If you need help to identify an organism, there are several great crowd-sourcing websites where expert advice can be sought. One of these is iSPOT, which has a very good track record of providing quick and accurate identifications. Even social media like *Twitter* can be used very effectively with careful choice of hashtags. For example, if it's a moth you would like identified try #teammoth or @ukmothidentification.

ONLINE TEST

Test yourself on identification at www.brightredbooks.net

FIELD TECHNIQUES FOR BIOLOGISTS: TAXONOMY

ONLINE

Use the link at www. brightredbooks.net to research a range of model organisms.

INTRODUCTION TO TAXONOMY

Taxonomy is the organisation of life into a hierarchy of groups of increasingly closely related species. These groups are known as domains, kingdoms, phyla, classes, orders, families, genera and species. If the taxonomic groupings are known for a particular organism, scientists can make predictions and inferences about its biology based on what is known about its relatives. The best-known species within each taxonomic group is considered to be the **model organism** for that group. This is likely to be the species within the group that is the easiest to keep, the easiest to study, the most useful or the most harmful!

The American robin (Turdus migratorius), shown left, can be easily identified by its reddish chest and belly. Taxonomically it is more closely related to the European blackbird than to the superficially similar Eurasian robin (Erathacus rubecula), shown right. The latter individual shown here was photographed in the Canary Islands and may, in fact, be a separate, but closely related, species to the Eurasian robin of Europe – studies of the degree of DNA-sequence divergence have so far proved inconclusive.

MODEL ORGANISMS

Model organisms are the **best-studied species within a taxonomic group**, or the best choice of organism for study of a particular biological process (such as cell division). As the members of a taxonomic group are related, the information from the study of a model organism can be applied to other species in that taxonomic group. This is advantageous as those other species may be more difficult to study directly. The following model organisms have been particularly important in the advancement of modern biology.

Escherichia coli

Escherichia coli is the most-studied member of the bacterial domain. Researchers have been able to culture *E. coli* in laboratories for many years; originally, it was used to study basic bacterial processes and more recently it has been used in biotechnology. For example, the use of restriction enzymes to genetically modify plasmids and transform cells was pioneered using *E. coli*. In addition, the interaction of inducers with repressor proteins in gene expression was first understood using this species.

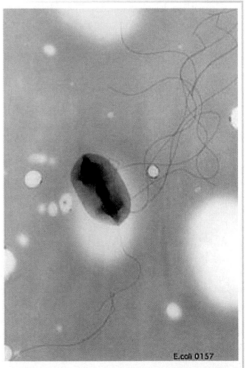

E.coli O157

An *E. coli* of the pathogenic form O157, which is capable of causing severe and sometimes fatal food poisoning. Non-pathogenic forms are common in the human gut. The pathogenic form has additional genes, including those for the shiga toxin protein, probably as a result of transformation by a virus vector.

contd

Arabidopsis thaliana

The flowering plant *Arabidopsis thaliana*, also known as thale cress, is a small rapidly growing plant of the mustard family. It was the first plant to have its entire genome sequenced, and its small genome and small number of chromosomes has made it a good model for the study of genomics, proteomics and metabolomics.

Caenorhabditis elegans

The roundworm, *C. elegans*, is the model organism for the abundant and diverse group that forms the Phylum Nematoda. This species was the first multicellular organism to have its entire genome sequenced. Much of our understanding of developmental biology, apoptosis and meiosis stems from research on this species. Many species of nematodes are parasitic, so research into this species also has the potential for applications in disease control.

The developmental pathway of each and every cell in a *C. elegans* roundworm is known and documented. Depending on whether it is a hermaphrodite or a male, a *C. elegans* worm consists of either 959 or 1031 somatic cells, there are no females in this species.

A picture of thale cress from an eighteenth century classification guide. Many mutant varieties were recognised and this led to its use as a model organism.

Drosophila melanogaster

The fruit fly is a model arthropod (phylum Arthropoda). Much of the research carried out using *Drosophila* has been related to genetics, mutation and evolution. We share 75% of our genes with the fruit fly and some of their cells have enlarged chromosomes that are very easy to study under the light microscope. This species of fruit fly is very easy to keep in the laboratory, has a short generation time and many different genetic markers that can be easily studied.

ONLINE

Check out the link at www.brightredbooks.net if you want to find out more about *C. elegans*.

Mice, rats and zebrafish

The chordates (phylum Chordata) mice, rats and zebrafish are among the most commonly used model vertebrates for the study of normal and abnormal physiology. Neurological and toxicological studies have been used extensively in all three species.

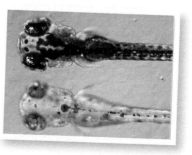

A transparent embryonic variety of the zebrafish is particularly useful in studies of heart development and function.

ONLINE TEST

Test yourself on taxonomy at www.brightredbooks.net

THINGS TO DO AND THINK ABOUT

In order to increase the power of the studies in different model organisms, there is now an attempt to ensure that a common language is used across different species to describe the various mutant and wild-type phenotypes. For an example, have a look at the first figure in the PLOS article at www.brightredbooks.net.

FIELD TECHNIQUES FOR BIOLOGISTS: THE TREE OF LIFE

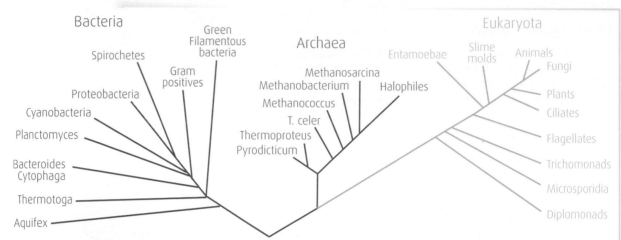

Genome comparison studies have shown life to be divided into three main domains, the Bacteria, the Archaea and the Eukaryota. Note that the Archaea are more closely related to the Eukaryota than they are to the Bacteria.

UNDERSTANDING PATTERNS OF EVOLUTION

The study of genomics and proteomics has revolutionised our understanding of the related nature of all life on earth. By comparing similarities and differences in the sequences of either proteins or DNA, it is possible to establish conclusively which organisms are more closely related to one another and which ones are distant relatives. Thousands of phylogenetic studies have now been combined to reveal the overall pattern of evolutionary relationships of the whole tree of life. The comparison of these sequences reveals that life has diverged into three domains: the **Archaea**, **Bacteria** and **Eukaryota**.

Patterns of descent are not always obvious when observing phenotypic characters alone. Evolution can produce closely related species that look very different to one another, and distantly related species that look very similar. The comparison of genetic evidence reveals relatedness that is obscured by this divergent or convergent evolution.

Divergent evolution

Divergent evolution is the development of **differing life forms from a common origin** and it results in closely related forms of life with very different phenotypic characteristics. This occurs when different selection pressures are acting on each lineage.

As a result, the ancestral characteristics are lost and are replaced by differing adaptive characteristics in the different lineages. Within the platyhelminths, the body plan of the parasitic tapeworms is quite different from their non-parasitic relatives. In lineages that have few competitors (maybe as a result of a mass extinction event, or in lineages that have had the good fortune to accumulate the mutations necessary for a particularly successful adaptation), divergent evolution can result in a radiation of many forms.

The molluscs provide an example of divergent evolution – it is not immediately obvious that a snail and an octopus belong in the same phylum.

This is a cactus isn't it? Actually, no – it is a Euphorbia from Africa. Cacti are from the Americas but these two families of flowering plants show a strong convergence as a result of exploiting similar niches.

DON'T FORGET

Molecular sequencing reveals relatedness.

ONLINE

Explore the tree of life online at www.brightredbooks.net

Convergent evolution

Convergent evolution is the **separate evolution of similar phenotypic adaptations** in lineages whose ancestors did not share these adaptations. This occurs when very similar selection pressures are acting on these unrelated lineages. For example, the structure of the eye in ourselves and in the octopus evolved independently in the vertebrates and the cephalopods. Likewise, the parasitic barnacle mentioned on page 81 is a member of one of two lineages of barnacles that have independently converged on this parasitic way of life. Convergent evolution is common; it seems that selection is efficient at arriving at a set of adaptations, even when starting off from different places.

OUR FAMILY TREE

There is now consensus amongst biologists on the relative positions of many phyla in the tree of life. Of course, some areas require more study and taxonomists do not agree on the position of every branch. However, there is overwhelming evidence from molecular sequences, from studies of anatomy, physiology and behaviour, and from fossils, that all life is related and has evolved into the diversity that we see today over the course of billions of years. It is no longer a credible scientific position to doubt this.

Bacteria

There are about 300 000 known types of bacteria. These single-celled organisms are thought to contribute more biomass than all animals and plants put together. We are familiar with bacteria for their key role that they play in nutrient cycles, their importance in modern biotechnology and the various pathogenic types. Examples of the latter include bacteria that cause bubonic plague, leprosy, anthrax, cholera and tuberculosis.

Many bacteria have had their entire genomes sequenced.

Archaea

These are single celled and differ from other forms of life in their ribosomal RNA structure and in the molecular structure of their lipids. The archaea are relatively poorly known – some are extremophiles; others have unusual metabolic activities, such as methanogenesis. There are about 4000 types, many of which are only known from genomic sequences sampled from the environment. There are no known parasitic or pathogenic members of this domain.

Eukaryota

There are over 2 million known species of eukaryotes, which have the shared cellular features of a true nucleus and membrane-covered organelles. The multicellular forms of life are found in this domain, but there are also many single-celled eukaryotes. Within eukaryotes, there are the familiar kingdoms of plants, fungi and animals. Of those three, the plant kingdom is the least closely related. It has major divisions such as mosses, liverworts, ferns, conifers and flowering plants.

Plant division: Moss

Plant division: Liverworts

Plant division: Ferns

Plant division: Conifers

Plant division:
Flowering plants

Plant divisions

Animal phylum: Chordata
(sea squirts and vertebrates)

Animal phylum:
Arththropoda (joint-legged invertebrates: segmented body typically with paired appendages)

Animal phylum: Nematoda
(round worms: very diverse, mainly parasitic)

Animal phylum:
Platyhelminthes (flatworms: bilateral symmetry, internal organs but no body cavity, many parasitic)

Animal phylum: Mollusca
(diverse, many with shells)

Animal divisions

THINGS TO DO AND THINK ABOUT

The comparison of the human genome with other species reveals remarkable similarities. Many gene sequences, such as the *Hox* genes that control developmental segmentation in animals as different as humans and *Drosophila*, have been conserved over long periods of evolutionary change.

ONLINE TEST

Head to www.brightredbooks.net to test yourself on the tree of life.

FIELD TECHNIQUES FOR BIOLOGISTS: MONITORING POPULATIONS

The monitoring of a population involves counting or estimating the number or density of a species on more than one occasion, not simply providing a one-off inventory of the species in an area. Instead, biologists aim to detect changes in populations over time. By doing so, they are more likely to be able to determine the natural phenomena or management actions that are responsible for any changes.

DON'T FORGET

Presence and absence are binary data, with a simple yes or no answer. Abundance has a scale, which may give truly continuous numerical data or which may be expressed in discrete categories such as abundant, common, frequent, occasional and rare.

INDICATOR SPECIES

Since each species tends to exist only within its particular niche (see page 78), the presence of a species tells us that the environmental conditions in the area meet its fundamental niche conditions. To gain information about a specific environmental condition, such as the presence of specific pollutant, we monitor the **presence**, **absence** or **abundance** of species that are either **intolerant** (sensitive to) or **tolerant** of (favoured by) that specific environmental condition. You will be familiar with the idea that different forms of lichen have different sensitivities to sulfur-dioxide pollution. If only the more tolerant forms are found, the area is experiencing relatively high levels of pollution. If the most sensitive forms are present, there must be very little sulfur-dioxide pollution in the local environment.

The proportion of drinking-water tests in Scotland that fail is declining thanks to improvements in sewage treatment facilities. The most frequent reason for failure is the presence of fecal coliforms in the water and these are an indicator of contamination with pathogens such as *Cryptosporidium*.

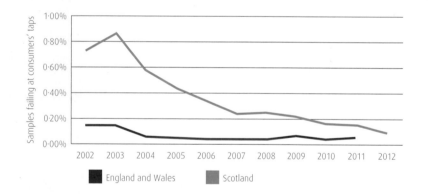

Freshwater sampling for indicators

You may be familiar with the idea of sampling freshwater invertebrates in order to assess river quality. High populations of mayflies indicate unpolluted water that is rich in dissolved oxygen. In contrast, high populations of bloodworms, which are tolerant to low oxygen levels, and low populations of the sensitive mayflies indicate that the waterbody is polluted with organic waste.

We sample freshwater supplies to look for indicator species of bacteria. The presence or absence of these coliforms indicates the likelihood of contamination with the Cryptosporidium parasite. This parasite uses the cells lining the small intestine to multiply and this affects the permeability of the gut wall. Cryptosporidiosis is the second biggest cause of diarrhoeal-disease mortality in infants.

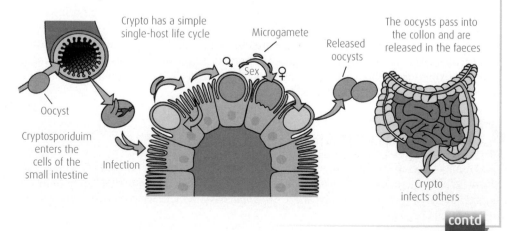

contd

Another indicator that is monitored in our rivers, beaches and drinking water supplies is the presence of fecal coliforms; in 2012, for example, over 64 000 fecal coliform tests were carried out on drinking water supplies in Scotland. Fecal coliforms are bacteria such as *E. coli* that originate from mammalian guts, and their presence in a water supply indicates contamination with sewage. More importantly, the presence of fecal coliforms indicates the possibility of the presence of potentially pathogenic organisms such as the human-gut parasite *Cryptosporidium*. Unacceptably high fecal coliform counts result in water from affected treatment works being diverted out of the drinking water supply until successful action has been taken.

METHODS OF MARKING

When monitoring populations, it is often helpful to be able to identify individual organisms. To do so, various methods of marking have been developed. Methods of marking include banding, tagging, surgical implantation, painting and hair clipping. The method of marking and subsequent observation must minimise the impact on the study species.

Method of marking	Description
Banding	A visible coded metal or plastic loop attached around part of the organism
Tagging	A visible coded metal or plastic tag inserted into or onto the organism
Surgical implantation	A microchip inserted into the organism; read by an electronic reader
Painting	A code painted onto the surface of an organism, such as a shelled mollusc
Hair clipping	A distinctive area of fur trimmed on a small mammal

This neck-ringed greylag goose was banded in Orkney and then sighted in Lothian. The unique mark allows the movement of this individual between these two sites to be detected. If enough individuals are marked, the movements of the population between breeding and wintering areas can be monitored.

MARK AND RECAPTURE

Mark and recapture is a method for estimating population size. A sample of the population is captured and marked (*M*), and released. After an interval of time, a second sample is captured (*C*). If some of the individuals in this second sample are recaptured (*R*) then the total population $N = (MC)/R$, assuming that all individuals have an equal chance of capture, that there is no immigration or emigration and that marked and released individuals can mix freely with the main population.

Example:

In this example, 26 greylag geese are caught and marked. The size of the population is unknown. Some time later, 21 geese are captured, of which 3 were marked. Work out the size of the total population using the formula $N = (MC)/R$. Once you have done that you could check your answer by counting the entire population.

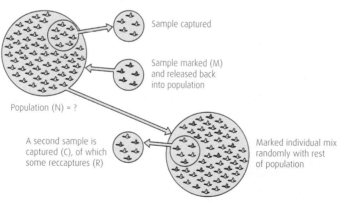

Sample captured

Sample marked (M) and released back into population

Population (N) = ?

A second sample is captured (C), of which some recaptures (R)

Marked individual mix randomly with rest of population

ONLINE

If you see a banded or tagged bird and know what species it is, then you can add to the global pool of scientific knowledge by submitting your sighting online. Find the link at www.brightredbooks.net

ONLINE TEST

Test yourself on monitoring populations at www.brightredbooks.net

THINGS TO DO AND THINK ABOUT

The Comma butterfly is spreading through Scotland at a rate of around 10 miles north per year as a result of climate change. Help to record the spread of this population by submitting any sightings to the iRecord (http://www.brc.ac.uk/irecord/) website.

FIELD TECHNIQUES FOR BIOLOGISTS: MEASURING AND RECORDING ANIMAL BEHAVIOUR

Behaviour is defined as the observable response that an organism makes to an internal or external stimulus.

Some of the most complex behaviours are shown by primates. Here baboons show (a) courtship through meat sharing in yellow baboons, (b) parental care in olive baboons and (c) learning through play in olive baboons.

STUDYING ANIMAL BEHAVIOUR

In school settings, animal-behaviour experiments are often restricted to invertebrate organisms, such as woodlice, and tend to reveal predictable responses to particular stimuli under laboratory conditions. Additionally, attempts are also made to quantify certain aspects of human behaviour, such as in demonstrating learning curves or simple tests of memory or processing speed.

At Advanced Higher level, it is important to recognise that behaviour is an extremely complex adaptation, and that its study involves variables that can be very difficult to identify, measure or control. The inherent variability in many behavioural responses means that **careful experimental design** has to be at the heart of this branch of biology. This variability also means that statistical methods are often used to highlight the significance (or otherwise) of experimental findings.

To help understand complex processes, scientists tend to start off by using a reductionist approach. In behaviour, this means the breaking down of complexity into defined and quantifiable categories. Once lots of conclusions have been drawn about many individual experiments and observations, these can synthesised into a coherent model and we move closer to understanding whole complex system.

Forms of quantitative data

Scientific analysis requires data that have been gathered in a **factual** and **objective** manner. The following categories of data are commonly used to quantify animal behaviour:

- latency – the interval of time between a stimulus and its response; for example, the time interval between the appearance of a female stickleback and the male starting his zig-zag dance

- frequency – how often a behaviour occurs; for example the number of zig-zags a male performs in his courtship dance

- duration – the length of time for which a behaviour occurs; for example, how long the zig-zag behaviour lasts.

contd

Anthropomorphism

Anthropomorphism is the attribution of human motivation, characteristics or behaviour to non-human animals. The use of anthropomorphism to draw conclusions about or to explain animal motivation and behaviour should be avoided. For example, it would be wrong to assume that a male stickleback is swimming in a zig-zag manner because he is either confused or forgetful. Only through objective study does it become apparent that the zig-zag behaviour is a sign-stimulus to the female to encourage her to follow the male to his nest.

The stickleback provides a classic study of courtship behaviour. Before studying courtship, an ethogram would have been made so that the researchers could be sure their study was repeatable and their findings comparable with those of other studies. In an ethogram, each behaviour is defined clearly.

ETHOGRAMS

An ethogram describes all the behaviours shown by a species in a wild context, allowing observation and recording of the amount of time spent exhibiting each type of behaviour. These time budgets allow comparisons of the frequency and durations of different behaviours to be made.

Stickleback courtship: the male is trembling against the female and this stimulates her to lay her eggs in his nest.

An ethogram suitable for the study of stickleback courtship.

Behaviour	Description
Foraging	Searching for food
Feeding	Grabbing food
Cannibalism	Feeding on eggs
Escaping	Rapid swimming ahead of other fish
Courtship: male dance	Swimming in a zig-zag pattern
Courtship: following	Female swimming slowly closely behind male
Courtship: indicating	Male holds body at angle and repeatedly touches nest entrance with mouth
Courtship: entering nest	Female entering nest
Courtship: trembling	Male repeatedly nudging female
Reproduction: egg laying	Female laying eggs
Reproduction: entering nest	Male entering nest
Parental care: nest-building	Male moving weed to nest site
Parental care: guarding	Stationary position in vicinity of nest site
Parental care: fanning	Male fanning nest with fins
Parental care: chasing	Aggression and rapid swimming behind other fish

VIDEO LINK

Watch the stickleback courtship at www. brightredbooks.net

THINGS TO DO AND THINK ABOUT

Do you find this image scary or appealing? It may seem strange, but your response to that question could be altered by a parasite, *Toxoplasma*. *Toxoplasma* is a protozoan parasite of cats. As it is a trophic-niche parasite, it returns to its definitive host through the food chain, such as in the body of a rat or a mouse. *Toxoplasma* is known to modify the behaviour of the intermediate host in ways that tend to increase the chances of successful consumption by a feline predator. For example, the degree of fear of and aversion to cats is lowered and infected rodents tend to choose to spend time in areas that have the scent of cat urine. Toxoplasma can also infect humans and the effects appear to parallel those in other intermediate hosts – personality traits associated with risk-taking behaviours have been shown to be more common in humans who are infected with the parasite. The mechanisms most probably responsible for these observed changes are an increase in dopamine, increase in testosterone in males and changes to the regulation of key genes in the amygdala, the part of the brain associated with emotional responses. Consider how you would design experiments that could be carried out to investigate the types of changes that occur in infected mice and humans.

ONLINE TEST

Test yourself on measuring and recording animal behaviour at www. brightredbooks.net

EVOLUTION BY DRIFT AND SELECTION

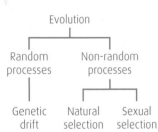

The causes of evolution

WHAT IS EVOLUTION?

Evolution does not happen to an individual organism in its lifetime. The organism may well change in response to environmental influences, but these changes cannot be passed on to the next generation. For example, bonsai trees are pruned and trained so that they grow small, but the seeds from bonsai trees can still produce full-height offspring.

Evolution happens to a **population** over a number of **generations**. The individuals in the population show genetic variation as they carry different versions of **inherited traits**. Evolution is seen when the **proportions of individuals** with different traits in the population change over a number of generations.

Mutations

Different versions of inherited traits have different DNA base sequences as a result of genetic **mutation**. If a mutation results in a base sequence which codes for a different amino acid in a polypeptide, then the mutation has produced a **novel allele** – a new version of the gene. Other mutations produce new combinations of alleles in the organisms.

There are three possible effects of mutations on the evolutionary fitness of an individual. A large number of mutations are **neutral**, because there is no effect on the evolutionary fitness of the individual. Some mutations are **harmful** because they reduce the individual's fitness. Very rarely, mutations are **beneficial** to the fitness of an individual.

WHAT CAUSES EVOLUTION?

Evolution occurs through the random process of genetic drift, and the non-random processes of natural selection and sexual selection. All of these mechanisms cause changes to the frequency of inherited traits in a population – the allele frequency – over a number of generations.

Genetic drift

Genetic drift is the **random** increase or decrease in frequency of inherited traits over a number of generations. There is no selection pressure on the alleles, so this form of evolution is particularly seen when the allele has a neutral effect on evolutionary fitness.

The change in allele frequency is down to chance events which mean some individuals pass on their inherited traits and some do not. These chance events could be in survival (e.g. some individuals survive a forest fire because wind blew the fire in a different direction) or in reproduction (e.g. there are only a few nest sites available).

Genetic drift has a greater effect in small populations. Any chance event will randomly affect a sample of the population and a small population is already a small sample. If 25% of a population is randomly selected to pass on

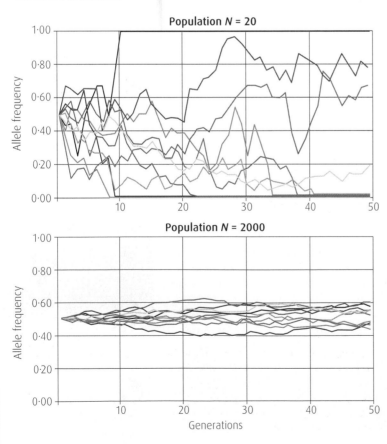

Effect of population size on genetic drift. Each graph shows ten simulations over 50 generations for a hypothetical allele starting at 0·5 frequency.

contd

its alleles, then a population of 20 only has 5 breeding individuals, while a population of 2000 still has 500 breeders. This means that alleles are more likely to be lost from the gene pool of a small population.

Natural selection

Natural selection is a **non-random** process which affects the allele frequency in the population over generations. It does this by selecting individuals based on their **survival chances** and thus increases their chances of passing on their inherited traits.

Alleles that are beneficial increase survival, so are more likely to be passed on to the next generation; their frequency will increase. Conversely, alleles which reduce evolutionary fitness reduce survival chances and are less likely to be passed on; they will decrease in frequency in subsequent generations.

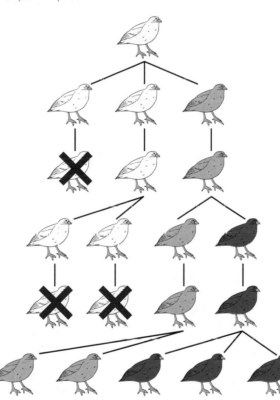

Organisms produce more offspring than the environment can support.

Individuals with variations that best fit their environment are more likely to survive and breed.

Favoured traits are inherited so they are more likely so become more frequent in subsequent generations.

A summary of the process of natural selection.

Sexual selection

Sexual selection is the **non-random** increase in the frequency of alleles that make mating and reproduction more likely. While this sounds similar to natural selection, the selected alleles do not increase survival chances. Some of the sexually selected alleles produce characteristics which may actually decrease the individual's survival chances, e.g. the peacock's long tail. However, if the individual survives, then the characteristic gives it an improved chance of reproducing, so the allele is more likely to be passed on; it will increase in frequency in subsequent generations.

Selection changes the allele frequency in subsequent generations. Here, the allele for darkness could be increasing survival chances of the individuals (natural selection), or increasing their reproductive chances (sexual selection).

DON'T FORGET

Natural selection works on alleles that increase survival chances. Sexual selection works on alleles that increase reproduction chances.

ONLINE

Read some good explanations of the role of chance in drift and selection at www.brightredbooks.net

VIDEO LINK

Watch all the features of natural selection being brilliantly demonstrated in a real population at www.brightredbooks.net

DON'T FORGET

Genetic drift is random. Natural selection and sexual selection are non-random.

VIDEO LINK

Watch an explanation of genetic drift at www.brightredbooks.net

VIDEO LINK

Test yourself on this topic at www.brightredbooks.net

THINGS TO DO AND THINK ABOUT

Changes in the phenotype proportions of a population over generations is not necessarily due to evolution. For example, there has been an increase in the proportion of obese individuals in UK population over last 25 years, but this is not an example of evolution! The population still has the same allele frequency, but the expression of traits has been changed due to environmental changes such as diet and lifestyle.

FITNESS AND THE RATE OF EVOLUTION

ABSOLUTE AND RELATIVE FITNESS

Evolutionary fitness can be considered in two different ways:

- **Absolute fitness** is the ratio of **frequencies of a particular genotype** in one generation compared to the **previous generation**.

- **Relative fitness** compares the **absolute fitness of one genotype** with the absolute fitness of the most **successful genotype**.

DON'T FORGET

Absolute fitness is calculated using the genotype frequencies, not the raw numbers.

In the diagram, the parent population has 10 individuals and only one has the genotype **dd**. This genotype has a frequency of 0·1 in the parent population. Since this genotype **dd** is found in 2 of the 8 offspring it has a frequency of 0·25 in the next generation. The absolute fitness compares the offspring genotype frequency with the previous generation so, for **dd** it is 0·25 : 0·1 which is 2·5 : 1. The absolute fitness of genotype **Dd** is 0·5/0·3, which is 1·7.

The relative fitness compares the absolute fitness. The genotype **dd** is most successful with an absolute fitness of 2·5 so its relative fitness is 2·5/2·5 which is 1·0. The relative fitness of **Dd** is 1·7/2·5 which is 0·7.

generation

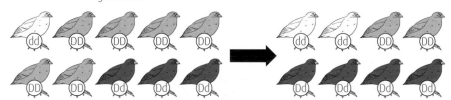

Absolute and relative fitness can be calculated from this diagram.

FACTORS THAT INCREASE THE RATE OF EVOLUTION

1. Higher selection pressures

High selection pressures increase the rate of evolution.

Selection pressures affect the heritability of traits. In natural selection, the selection pressures affect the individual's survival chances, while in sexual selection the pressures affect the individual's reproductive chances.

Selection increases the heritability of beneficial traits and reduces the heritability of harmful traits. When there are **higher selection pressures**, the traits are being more favoured (or weeded out more) than if there are smaller selection pressures, so the rate of evolution will be **more rapid**.

2. Shorter generation times

If generation times are short, there are more generations in a set period of time. Since evolution happens over generations, having more generations provides more opportunities for selection pressures to operate and change the frequency of traits in the population.

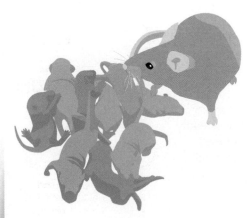

Short generation times and many offspring increase the rate of evolution.

For example, one female mouse can have between 5 and 10 litters each year, whereas a female elephant has just one offspring every 3 to 5 years. This means that the frequency of traits in the next generation can change every few months in mice and only every few years in elephants.

contd

The high reproductive rate of mice also means that they produce lots of gametes by meiosis, so they have a lot of DNA replication. This means there is a greater chance of mutation occurring, providing more alleles for selection and so increasing the rate of evolution.

3. Warmer environments

Most organisms do not regulate their body temperature, which remains the same as their environment. When the environment gets warmer, all the enzymes in the organism work faster. Amongst other things, the germ cells which produce gametes by meiosis divide more frequently and so there is more DNA replication, which means a greater chance of mutation.

This effect has also been found in mammals that are homeothermic and maintain relatively constant body temperatures. A mammal in a colder climate may hibernate or have periods of resting to conserve energy, because food is less available. They have a lower average metabolic rate across a year so the germ cells which produce gametes by meiosis divide slower, have less DNA replication and a lower chance of mutation.

4. Sexual reproduction

Combining genetic material from two different individuals during sexual reproduction increases the variation shown by the offspring. Greater variation leads to more selection and a faster rate of evolution.

Sexual reproduction also means that beneficial DNA sequences are shared between different lineages, producing offspring with new combinations of beneficial alleles. This is particularly important in the Red Queen Hypothesis and the co-evolution of hosts and parasites.

5. Horizontal gene transfer

Some organisms, such as bacteria, can pass genetic material to other members of their own species, or to other species. This is called **horizontal gene transfer** because the genetic material is not passing 'down' the generations by reproduction (vertical gene transfer).

Horizontal gene transfer allows the **sharing of beneficial DNA sequences** between different lineages of bacteria, using plasmids to transfer genetic material. This produces rapid evolution of a population because it quickly increases the frequency of a beneficial sequence (e.g. antibiotic resistance in bacteria).

Warmer temperatures increase the rate of evolution.

Sexual reproduction increases the rate of evolution.

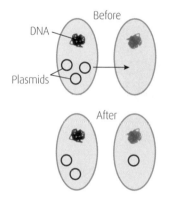

Before

DNA

Plasmids

After

Horizontal gene transfer increases the rate of evolution.

DON'T FORGET

There are five ways to increase the rate of evolution: higher selection pressures, shorter generation times, warmer environments, sexual reproduction, and horizontal gene transfer.

ONLINE

Read about the discovery of the effect of warmer temperatures on the rate of evolution in mammals at www.brightredbooks.net

VIDEO LINK

Watch the clip at www.brightredbooks.net about how selection increased the frequency of milk tolerance in humans.

ONLINE TEST

Test yourself on fitness and the rate of evolution at www.brightredbooks.net

THINGS TO DO AND THINK ABOUT

The common buzzard (*Buteo buteo*) has three colour forms caused by different genotypes. These are **DD** which has dark plumage, **dd** which has pale plumage and **Dd** which is intermediate. The intermediate colour form **Dd** has the best survival chances but the offspring of two **Dd** birds will have some **DD** and **dd** individuals. This means that **Dd** has a low absolute fitness in the next generation.

Two buzzards with the intermediate colour form.

CO-EVOLUTION

CO-EVOLUTION INVOLVES TWO SPECIES

DON'T FORGET

Co-evolution can be seen in pairs of species that interact frequently or closely.

Organisms are exposed to selection pressures and these pressures cause the evolution of the population over generations. If there are two species which have **frequent or close interactions**, they may show co-evolution. This is when a change in the traits of one species acts as a natural selection pressure on the other species. This is a 'tit-for-tat' process – each species is constantly evolving in response to the changes in the other species.

1. Some individuals of Species A have a trait that gives them a selective advantage in their interaction with Species B.

2. The proportion of individuals of Species A with the beneficial trait increases in the population over generations.

3. Some individuals of Species B have a trait that now gives them a selective advantage in their interaction with Species A.

4. The proportion of individuals of Species B with the beneficial trait increases in the population over generations.

One round of co-evolution. Each round can be started by new traits in either species.

Alice and the Red Queen run faster and faster, but stay in the same place.

The Red Queen Hypothesis

This hypothesis explains the idea of co-evolution using the analogy of the Red Queen from Lewis Carroll's *Through The Looking Glass*. In this book, the Red Queen runs faster and faster and Alice chases her but, when they stop running, Alice and the Red Queen are still in the same place. The Red Queen Hypothesis of co-evolution states that both species must 'keep running in order to stay still' in evolutionary terms.

The variation for selection can be produced by mutation and by sexual reproduction. Indeed, the Red Queen Hypothesis may explain the evolution of sexual reproduction as a way to produce variation more readily than by relying on mutation alone.

DON'T FORGET

The Red Queen Hypothesis means that both species are constantly evolving in response to changes in the other species.

ONLINE

Read more about the Red Queen Hypothesis at www.brightredbooks.net

EXAMPLES OF CO-EVOLUTION

Co-evolution can be seen in a number of ecological interactions.

1. Herbivores and plants

Plants have defences against being eaten (e.g. thorns, toxins) and herbivores have evolved ways to feed on these plants without coming to harm. As such co-evolution progressed, it led to some herbivores becoming so specialised that they are now exclusive to their food source.

Toxins found in Brassica plants protect against predation by most herbivores, but caterpillars of cabbage white butterflies can break down the toxins.

Cabbage white caterpillars feeding on a leaf of the Brassica rocket.

contd

2. Plants and their pollinators

Many species of angiosperms have co-evolved with specific pollinators. One part of this co-evolution occurs when an animal species evolves features to get nectar more effectively from one species of plant, and selection favours the flowers which are able to transfer the pollen to the animal more effectively.

It was Charles Darwin who first understood this relationship. He predicted the existence of a moth with a 28 cm proboscis, based on observations of an orchid which produces scent at night (when moths are active) and has nectaries at the end of a 28 cm flower tube. The predicted moth was discovered 20 years after Darwin died.

Darwin's orchid. The nectaries are at end of the long flower tube that reaches to the bottom of the picture.

3. Predators and their prey

Predation provides a strong selection pressure on prey populations, so co-evolution is evident in predator–prey relationships. Natural selection favours any trait which gives a prey animal a better chance of evading capture. In turn, this acts as a selection pressure for any trait which helps the predator to increase its chances of capturing its prey.

A zebra has a long stride to allow faster escape from predators.

This can be seen in the evolution of the horse family, which has developed longer limbs to allow faster running. Its ancestors 50 million years ago had feet with four toes and, by 3 million years ago, the horse was walking on the fingernail of its middle digit. Their predators (e.g. lions, wolves) have lengthened their stride by walking on the pads of their digits and evolving a flexible back.

4. Parasites and their hosts

Hosts and parasites are co-evolved, so parasites are specific to their host species. As predicted by the Red Queen Hypothesis, the evolution of new features in one species causes evolutionary changes in the other species, so both species are able to survive effectively.

Hosts that are better able to **resist and tolerate parasitism** have greater fitness, so this trait increases in the subsequent generations. Parasites are selected to be better able to **feed and reproduce, and to be transmitted to new hosts**. The last part of this unit looks in detail at parasite transmission, host defences and the ways in which parasites evade the host's defences.

The body louse has co-evolved with its human host. It lives in clothing and it emerges to feed on the body. The adults die at room temperature, so transmission is by contact between hosts.

DON'T FORGET

The ongoing co-evolution of hosts and parasites is a good example of the Red Queen Hypothesis.

VIDEO LINK

A nice low-tech animation of the Red Queen Hypothesis can be found at www.brightredbooks.net

THINGS TO DO AND THINK ABOUT

The human head louse and body louse are subspecies which feed on different parts of the body, and demonstrate resource partitioning. Phylogenetic studies estimate that the body louse diverged from the human head louse about 100 000 years ago, so this hints at the time when our ancestors started to wear some form of clothing.

ONLINE TEST

Test yourself on this topic at www.brightredbooks.net

SEXUAL AND ASEXUAL REPRODUCTION

COSTS AND BENEFITS

Evolution favours the individuals that are able to produce the maximum number of surviving offspring using least amount of energy. The simplest way to do this is by using asexual reproduction and it is clear that this evolved long before the more complex cellular processes needed for sexual reproduction. Both forms of reproduction occur today, so how do they compare in their cost effectiveness?

ONLINE

Read about the clever studies that have tested some of the hypotheses at www.brightredbooks.net

DON'T FORGET

It is not strictly true that only half of a sexually reproducing population produces offspring. See page 72 for more information on this.

Asexual reproduction	Sexual reproduction
• Metabolic costs are **lower** because it is based on the normal cell cycle and produces many identical offspring.	• Metabolic costs are **higher** since gametes are made by a special type of cell division and many fail to achieve fertilisation.
• The production of offspring can be **rapid** because the process is relatively simple and every member of the population can reproduce.	• Only half the population – the females – can actually produce offspring, so the reproduction rate is much **slower**.

So, since asexual reproduction seems to be the cost effective way to maximise offspring production, why is sexual reproduction the main strategy in virtually all complex organisms?

THE PARADOX OF SEX

More than 25 different hypotheses have been put forward to try to explain why sexual reproduction is so prevalent. The main stumbling block for them all is in trying to justify the existence of males! The disadvantages of needing males are that:

1. males are unable to produce offspring, so only **half the population** can actually reproduce; this reduces the reproductive effectiveness of the population as a whole

2. combining the genetic material of two parents to make new offspring means that only **half of each parent's genome** is passed onto offspring, so this **disrupts successful parental genomes**.

Some hypotheses suggest that the serious disadvantages of having males are outweighed by the benefits through sexual reproduction of an **increase in genetic variation** in the population. This genetic variation provides the raw material required for natural selection so that organisms can **continue adapting** to changing environments or to the ever changing parasite–host balance.

Other hypotheses propose that sex removes deleterious mutations from the gene pool. Indeed, there is currently no general explanation for the existence of sexual reproduction that is applicable to all eukaryotes.

Why do males exist?

DON'T FORGET

Genetic variation is the essential raw material for natural selection.

ASEXUAL REPRODUCTION CAN BE A SUCCESSFUL STRATEGY

Asexual reproduction in prokaryotes

Bacteria and archaea principally use asexual reproduction – they simply divide to produce new cells. Many of these organisms also have mechanisms which allow genes to be moved between individuals of the same generation; this is called **horizontal gene transfer**. Examples of these mechanisms include the **plasmids of bacteria and yeast**.

 contd

These allow the exchange of genetic material and so help to increase the rate of evolution, as is seen in the rapid spread of antibiotic resistance in bacteria.

Asexual reproduction in plants

Eukaryotes principally use sexual reproduction, though many plants can also use asexual reproduction. Flowering plants use their flowers for sexual reproduction but a wide range of methods of **vegetative cloning** has also evolved in flowering plants, including bulbs (daffodils, onion), corms (crocus, asparagus), tubers (potato, dahlia), stolons (strawberry, spider plants) and rhizomes (ginger, iris). Fungi and non-flowering plants (such as mosses and ferns) use asexually produced spores for dispersal to colonise new habitats.

Female aphid clones.

Asexual reproduction in animals

Asexual reproduction is much less common in animals. A few animal species use a process known as **parthenogenesis** (Greek: 'virgin creation') to produce offspring asexually, without needing males or **fertilisation**. Aphids and stick insects are examples of insects that produce multiple female clones, but can also produce males by deleting a sex chromosome.

One species of whiptail lizard, *Cnemidophorus murinus*.

Asexual reproduction is extremely rare in vertebrates. Some reptiles like whiptailed lizards and a handful of gecko species exist only as all-female populations that reproduce exclusively by parthenogenesis. These species are all the result of hybridisations between sexual species and they use modified versions of meiosis to produce diploid eggs that develop into female clones.

Garlic cloves are individual bulbs. Each can give rise to a new plant.

Parthenogenesis, although rare, is more commonly found in areas with **lower parasitism**. These include **cooler climates** that are disadvantageous to parasites and regions where there is a **low parasite density or diversity**. In these areas the Red Queen's race is being run more slowly, so there is a lower selection pressure on producing offspring with genetic variation.

CONDITIONS FAVOURING ASEXUAL REPRODUCTION

Using asexual reproduction to produce identical offspring can be beneficial when an organism is well adapted to a **very narrow and stable niche**. For example, aphids use asexual reproduction to quickly exploit the food source provided by new leaf growth, because they have a narrow feeding niche (piercing the phloem of plants) and live in a relatively stable microclimate (amongst plant leaves). Lawns of grass are very stable niches with specific conditions and daisy plants, *Bellis perennis*, use stolons to spread clones across the area.

Asexual reproduction is also be useful when **recolonising disturbed habitats**. Buttercup plants, *Ranunculus repens*, use stolons to spread into flower beds with clear soil and gain a competitive advantage by dominating the area before other plants can colonise with seeds.

THINGS TO DO AND THINK ABOUT

The natural variation generated by sexual reproduction has a big influence on experimental design in the life sciences. Biologists have to take account of this variation when sampling a population and analysing the data to make sure that they can distinguish this variation 'noise' from any experimental result or 'signal'.

DON'T FORGET

Parasites, and most of their hosts and vectors, are thermoconformers so all will grow and reproduce slower in cooler climates.

VIDEO LINK

Find out why aphids use parthenogenesis and then switch to sexual reproduction by watching the clip at www.brightredbooks.net

ONLINE TEST

Test yourself on sexual and asexual reproduction at www.brightredbooks.net

MEIOSIS PRODUCES HAPLOID CELLS

Organism	Diploid number of chromosomes
Fruit fly, *Drosophila melanogaster*	8
Thale cress, *Arabidopsis thaliana*	10
Round worm, *Caenorhabditis elegans*	12 (hermaphrodites) or 11 (males)
Yeast, *Saccharomyces cerevisiae*	32
Pig, *Sus scrofa*	38
Mouse, *Mus musculus*	40
Rat, *Rattus norvegicus*	42
Zebra fish, *Danio rerio*	54

Diploid numbers of some model organisms used in research.

DON'T FORGET

Sexual reproduction increases variation because it involves crossing over, independent assortment and combining haploid genomes from two individuals.

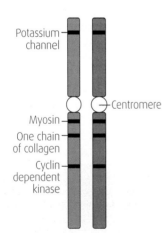

A diagram of the homologous pair of human chromosome 12, which has over 1600 genes. The four genes shown code for proteins that are covered elsewhere in the AH Biology course.

DIPLOID AND HAPLOID

The body cells of most eukaryotic organisms have an **even number** of chromosomes because they inherit one set of chromosomes from the female parent and one set from the male parent. All body cells are **diploid** – they have two sets of chromosomes – while gametes carry only one set of chromosomes and are **haploid**. The diploid number for humans is 46, which is two sets of 23 chromosomes.

Homologous chromosomes

The chromosomes in each set have a matching **homologous chromosome** that is the **same size**, has the **same centromere position** and has the **same genes at the same loci** (Latin: 'places', as in 'location'). Though the genes are the same on the homologous chromosomes, the alleles may be different as each chromosome in the pair has been inherited from a different parent.

Sexual reproduction combines haploid genomes

Fertilisation occurs when the haploid nuclei of two gametes fuse together to form a new diploid nucleus. The **combining of haploid genomes** from two different individuals produces a new combination of alleles in the offspring, so variation is increased. The production of haploid gametes starts with a type of cell division called **meiosis**.

MEIOSIS HAS TWO ROUNDS OF CELL DIVISION

Meiotic cell division starts with the interphase stages of the normal cell cycle. During the S-phase, the homologous chromosomes duplicate. This is still a **diploid cell** with two sets of homologous chromosomes; since each chromosome has been duplicated, it has four sets of genetic information. Meiosis takes this **gamete mother cell** through **two rounds of cell division** which separate the four sets of genetic information and so produce four haploid cells. Some or all of these haploid cells can then become gametes.

Meiosis

Meiosis I has two mechanisms to ensure that the gametes produced by one individual are highly genetically variable: **crossing over** and **independent assortment**. Meiosis II follows the same pattern as mitotic cell division and reduces the quantity of genetic material in the cells.

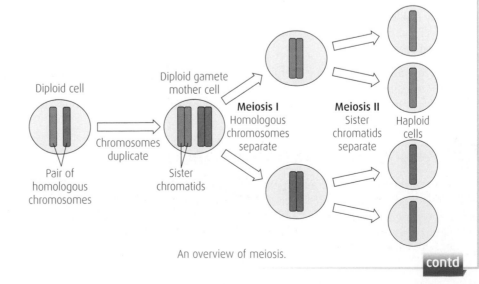

An overview of meiosis.

contd

Gamete formation

In human males, meiosis only starts at puberty and all the haploid cells develop into sperm cells.

In human females, it is much more complex. When she is born, each female already has about a million gamete mother cells at the chiasmata stage of meiosis I. After puberty, some of these cells are stimulated to divide further during each menstrual cycle, though only one will reach the chromosome alignment stage of meiosis II before stopping again. The division of meiosis I was asymmetric, so the future ovum is large and the other cell is much smaller. This means that the future ovum has a large amount of cytoplasm to sustain the early embryo, if the ovum is fertilised. The future ovum only completes meiosis II after it has been penetrated by a sperm cell. Again, the division is asymmetric and it is the haploid nucleus of the larger cell that fuses with the sperm nucleus.

VIDEO LINK

Meiosis is a dynamic process so an animation is a good way to really understand what happens. Check out the clip at www.brightredbooks.net

ONLINE

Find out more about the genetic disorders associated with each human chromosome

DON'T FORGET

The diploid or haploid number describes the **number of visible chromosomes**, not the number of copies of genetic information.

At the end of meiosis II, four genetically different haploid cells have been produced.

At the start of meiosis I, the chromosomes coil up and become visible. Each chromosome (Greek: 'coloured body') can be seen with two sister chromatids linked by the centromeres. For simplicity, our diagram shows just two homologous pairs.

The homologous chromosomes pair up so that they are aligned, gene by gene, to form a bivalent.

Chiasmata (Greek: 'cross marks') form at random positions between the homologous pairs and these allow the crossing over of sections of DNA between homologous chromosomes.

The nuclear membrane breaks down so that microtubules from the centrosomes (see page 41) can connect with the centromeres. The microtubules form spindle fibres linking across the cell and this aligns the homologous chromosomes at the equator, a plane across the middle of the cell. The chromosomes are positioned randomly, irrespective of the parental origin of the chromosomes; this is described as independent assortment. For simplicity, our diagram shows the maternal chromosomes (red) at one side and the paternal chromosomes (blue) at the other.

The microtubules of the spindle fibres shorten and pull on the centromeres so the homologous chromosomes separate to opposite ends of the cell.

A nuclear membrane forms around chromosomes and cytokinesis separates the two cells.

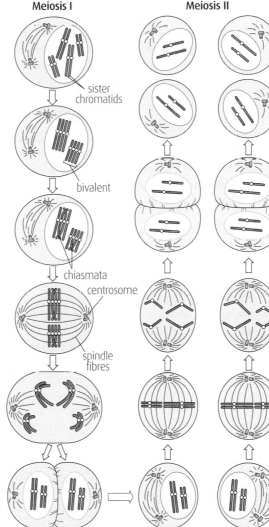

Meiosis I   Meiosis II

sister chromatids

bivalent

chiasmata
centrosome

spindle fibres

The new chromosomes group in each end of the cell and a nuclear membrane forms around them; cytokinesis separates the cells.

The microtubules shorten and pull on the centromeres, so the sister chromatids separate to opposite ends of the cell. Immediately after being separated, sister chromatids are called chromosomes (because they can be seen as separate 'coloured bodies').

The nuclear membrane breaks down. Centrosomes again send out microtubules and bind to the centromeres of each sister chromatid. The spindle fibres align the chromosomes across the equators of the cells.

Each cell is haploid, with just one copy of each homologous chromosome (though each cell still has two sets of genetic information). Notice that the sister chromatids are no longer identical, due to the crossing over.

THINGS TO DO AND THINK ABOUT

In 2006 a female Komodo dragon at Chester Zoo laid some viable eggs, even though she had never had any contact with a male. This was due to the meiotic haploid cells duplicating to produce diploid cells, which went on to develop. Only the eggs with male embryos hatched – to find out why go to http://www.scientificamerican.com/article/strange-but-true-komodo-d/

ONLINE TEST

Head to www.brightredbooks.net and test yourself on this topic.

MEIOSIS AND VARIATION

DON'T FORGET

Independent assortment produces new combinations of the maternal and the paternal chromosomes in the gametes.

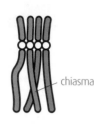

Chiasmata can only form between chromatids of two homologous chromosomes, never between chromatids of the same homologue

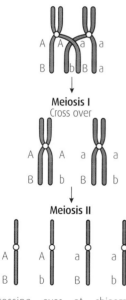

Crossing over at chiasmata produces recombination of the alleles of linked genes in the gametes. If there was no crossing over, the chromosomes would only have AB or ab as possible allele combinations.

VIDEO LINK

Watch a narrated animation of the effect of crossing over on combinations of alleles on chromosomes at www.brightredbooks.net

PRODUCING HAPLOID CELLS WITH VARIATION

Meiosis uses two different processes to produce cells that are genetically variable. Fertilisation then combines the genetic material from two different parents, so the range of possible variation in the offspring is huge.

Independent assortment

When homologous chromosomes are positioned on the equator during meiosis I, there is no control over which chromosome of each pair goes to which side of the cell. They are aligned **irrespective of their maternal or paternal origin**. This is called **independent assortment** because each pair is positioned randomly, with no dependence on the way that any other pair has lined up. When these **homologous chromosomes are separated** there are many possible combinations of chromosomes in the resulting haploid cells.

As shown in the diagram, even with just three pairs of homologous chromosomes, there are four possible alignments during meiosis I. When these three homologous pairs are separated by meiosis I and II, this results in $2^3 = 8$ possible combinations of chromosomes in the haploid cells.

The simple process of independent assortment leads to each gamete having a completely random selection of every homologous chromosome, producing **genetic variation** in the gametes. In humans, with 23 pairs, there are $2^{23} = 8\,388\,308$ combinations of chromosomes in the haploid cells.

Crossing over

The chromosomes of multicellular organisms have thousands of genes. For example, *Arabidopsis thaliana* has 25 498 genes on its five chromosomes, and *Drosophila melanogaster* has 15 682 genes on its four chromosomes. From this, it is obvious that many genes must be located on each chromosome and these are called **linked genes**. If independent assortment was all that happened during meiosis, the chromosomes in the gametes would have exactly the same alleles as the chromosomes that were inherited from the parents. **Chiasmata** allow the chromosomes to exchange DNA with their homologous partner. The homologous chromosomes, therefore, carry some different alleles from each other; the exchange means that chromosomes in the gametes can have **new combinations of alleles**. Unsurprisingly, this process is called **recombination**.

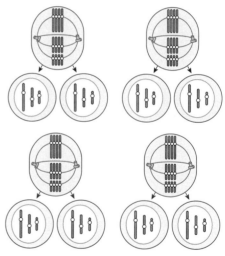

Independent assortment leads to variable gametes.

FERTILISATION DURING SEXUAL REPRODUCTION

In summary, there are three processes that help sexually reproducing organisms to produce genetically variable offspring. **Independent assortment** and **crossing over** during meiosis produce variable gametes. **Fertilisation** during sexual reproduction then brings the genetic material from two different parents together in one organism.

Even without crossing over, one pair of human parents could produce over 70 million million possible diploid chromosome combinations. Adding in the effect of crossing over means that the number of different possible allele combination is astronomical. So, except for identical twins, each of us is truly genetically unique in the whole history and future of life on Earth.

LINKAGE MAPPING

Chiasmata form at **random positions** on the chromosomes. This means that the further apart genes are on a chromosome, the more likely they are to have a chiasma form between them, leading to new allele combinations in the gametes.

Imagine we were studying an organism with the chromosome shown here. A cross between organisms with genotype **AABB** and **aabb** would result in heterozygous offspring with chromosomes carrying **AB** and **ab**. If this heterozygote **ABab** is then crossed with an **aabb** organism, then all the recombinant genotypes will show in the phenotypes of the offspring.

The percentage of recombinants in the offspring of this cross gives an indication of how far apart the genes are on the chromosome. The further apart the genes are, the more likely it is for a chiasma to form between them, giving a higher proportion of recombinant offspring. This means that the percentage recombination can be used to map the distance between the genes. If the cross produced 20 recombinant offspring out of a total of 200 offspring, the percentage of recombinants is 10%, a map distance of 10.

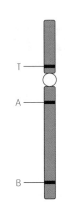

Chiasma are more likely between gene A and B than between A and T.

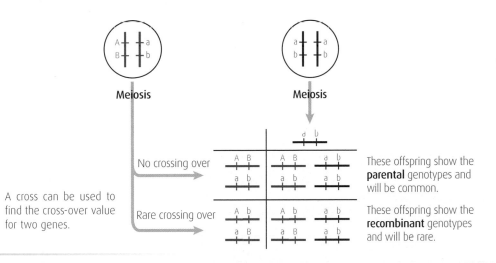

A cross can be used to find the cross-over value for two genes.

These offspring show the **parental** genotypes and will be common.

These offspring show the **recombinant** genotypes and will be rare.

DON'T FORGET

Chiasmata are the structures that allow crossing over. Crossing over produces new combinations of alleles on the chromosomes.

SOME LIFECYCLES USE MEIOSIS TO FORM HAPLOID ORGANISMS

Many lifecycles have a diploid organism which uses meiosis to produce haploid cells that **form gametes directly**; these can combine with gametes from another parent to produce a new diploid organism. In other organisms, the haploid cells produced by meiosis immediately undergo mitosis to form a **haploid organism** that can live independently; this haploid organism forms gametes at a later stage by **cell differentiation** and these gametes take part in fertilisation to produce a diploid organism. For example *Plasmodium sp.*, which cause malaria, are haploid in humans and only become diploid in the mosquito.

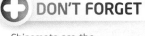

VIDEO LINK

Check out the animation at www.brightredbooks.net which explains the ideas of crossing over and map distance really well.

THINGS TO DO AND THINK ABOUT

You should be able to work a genetic linkage map if you are given the percentage recombinants for some crosses.

Gene pair	A × B	A × T	B × T	B × Q	Q × T
Percentage recombinants	10	5	15	3	12

Go for the largest value first, so **B** and **T** are furthest apart. **A** fits in between **B** and **T** (as it is 10 from **B**, and 5 from **T**). Similarly, **Q** fits between **B** and **T** (as **Q** is 3 from **B**, and 12 from **T**). The gene order should **B-Q-A-T**. Draw out the map and then work out why the expected percentage of recombinants for **A × Q** should be 7%.

ONLINE TEST

Test your knowledge of this topic at www. brightredbooks.net

SEX DETERMINATION

We are most familiar with species that have two separate sexes, male and female. The sex of an individual can be determined by **genetic factors** or by **environmental factors**.

There are, however, many species that are **hermaphroditic** which means that each individual has both male and female reproductive structures. There are usually mechanisms that mean individuals cannot fertilise themselves. For example, many flowering plants have male stamens and female carpels in the same flower, though they are not usually active at the same time. Land snails are hermaphrodites, so two individuals can entwine their bodies and be fertilised by each other.

DON'T FORGET

Remember that each individual has a full double set of all chromosomes making up the full chromosome complement. So a human female has cells with **XX and 44** other chromosomes.

SEX CAN BE DETERMINED BY GENETIC FACTORS

Sex chromosomes

Live-bearing mammals and some insects (including *Drosophila*) have chromosomes that determine the sex of the individual. These organisms are diploid, so have two full sets of homologous chromosomes. However, one pair of chromosomes is an exception to the normal definition of a homologous pair (see page 68) and these are the **sex chromosomes**, called X and Y in live-bearing mammals. Females have a homologous pair of X chromosomes (XX) while males have one X and a smaller Y chromosome (XY). During meiosis in the male, the X and Y chromosomes pair up because there is a small area of homologous genes near the centromere. Meiosis ensures that there is one sex chromosome in each gamete, so all the gametes from an XX female carry an X chromosome (this the **homogametic** sex). For the XY male, half the gametes carry an X chromosome and the other half carry a Y chromosome (males are the **heterogametic** sex).

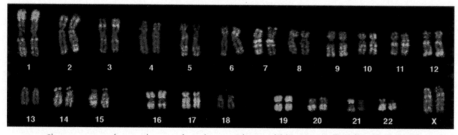

Chromosomes from a human female. A male would have X and Y chromosomes.

Sex chromosomes have also evolved in birds and some reptiles. In these animals, the homogametic sex is the male (ZZ) and the heterogametic sex is the female (ZW).

In the live-bearing mammals, it is a **single gene on the Y chromosome** (called SRY) which causes the embryo to develop as male. The SRY gene is thought to act as a master switch which triggers a cascade to activate all the male genes in the genome. If there is no SRY gene, then the default pathway causes the embryo to develop as female.

Sex linkage

The X chromosome has many genes which do not have homologous alleles on the smaller Y chromosome and this leads to **sex-linked patterns of inheritance**. The inheritance of white eyes in Drosophila was the first characteristic in which this sex-linked pattern was worked out. In this case, the dominant allele **R** gives red eyes, while the recessive allele **r** gives white eyes.

In females, a recessive allele **r** on one X chromosome can be masked by a dominant allele **R** on the other X chromosome, giving the red-eyed phenotype. The heterozygous genotype is shown as $X^R X^r$ and individuals with this genotype are described as **carrier females**. A

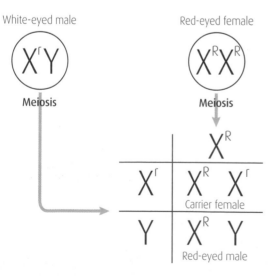

The first sex-linked cross for Drosophila.

contd

male with a recessive **r** on the X chromosome will be shown as X^rY and will have the recessive white-eyed phenotype, because the smaller Y chromosome has no homologous gene to mask the effect.

X-inactivation

Male cells have only one X chromosome and female cells have two. You might think that female cells would produce twice as much of the proteins from the genes on their two X chromosomes, but they don't.

During early embryonic development in females, **most of the genes** on one X chromosome in each cell are **inactivated** so the cells have a single working copy of the X chromosome genes. There are a few homologous genes shared by the X and Y chromosomes, so these have to be left active on the inactivated X chromosome. This X-inactivation is an example of **dosage compensation** which ensures that the females have exactly the same level of gene products as a male and do not get a **double dose of gene products** as this could be harmful to the cells.

A tortoiseshell cat has the genotype X^BX^b. Random X-inactivation means that some cells have X^B and show black fur, while other cells have X^b and show ginger fur.

The phenotype of a male always shows the effect of the genes that are inherited on the X chromosome. Females who inherit a deleterious allele (whether dominant or recessive) on one of their X chromosomes are less likely to be affected by its effects. Because X-chromosome **inactivation is random** about half of the cells in any tissue will have a working copy of the gene.

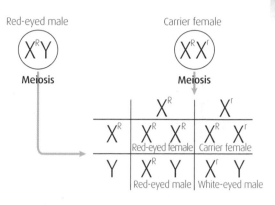

The second sex linked cross for Drosophila, using the offspring from the first cross.

SEX AND SEX RATIO CAN BE DETERMINED BY ENVIRONMENTAL FACTORS

There are environmental factors which can change the sex of embryos or adults, or the sex ratio of the offspring produced by the parents.

Environmental factor	Example	Effect on sex, or sex ratio
Temperature	Hermann's tortoise	Incubation of eggs at less than 31°C produces all male offspring and above 32°C gives all females.
Size	Clown fish	These live in groups with one large female and some smaller females. When the large female is removed, the largest male becomes a female.
Competition	Lesser mouse lemur	If a solitary female detects urine of another female, she produces more male offspring.
Parasitic infection	Insects infected by *Wolbachia* bacteria	Bacterial infection of the eggs kills the males or feminises them.

THINGS TO DO AND THINK ABOUT

Occasionally a gamete carries two X chromosomes instead of just one, due to spindle failure during meiosis. This leads to offspring with an extra sex chromosome, so they could be XXX or XXY. A person with XXX will develop as a female and will have two of her X chromosomes inactivated. A person with XXY will develop as a male because of the SRY gene, and they will have one inactivated X chromosome.

DON'T FORGET

Sex linkage is due to genes on the X chromosome.

DON'T FORGET

Either X chromosome can be inactivated in a cell – inactivation is random.

ONLINE

Find out more about sex linkage at www.brightredbooks.net

VIDEO LINK

An explanation of the sex-linked inheritance of haemophilia in humans can be seen at www.brightredbooks.net

ONLINE TEST

Revise this topic by taking the test at www.brightredbooks.net

PARENTAL INVESTMENT

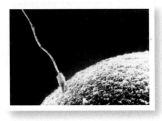

Parental investment can include parental care. (a) Mammals invest in parental care – this vervet monkey mother is providing resources directly from her body when suckling a youngster. (b) Parental care also involves the protection of young. Here a black-naped tern attacks a white-bellied sea-eagle many times its own size – a risk it will only take in the protection of its young.

The difference in size between the sperm and the egg is shown clearly in the picture of mammalian fertilisation. The egg is estimated to have 100 000 times the volume of the sperm and is, therefore, a much more costly cell to produce.

DON'T FORGET ✚

There is far greater parental investment involved in the production of an egg by a female than in a sperm by a male parent.

INTRODUCTION

Parental investment is the use of resources by a parent to benefit future or existing offspring. Investing resources in offspring is obviously **costly** to the parent. The **benefits** are that it can result in increased numbers of offspring and their improved chance of survival, thereby increasing the evolutionary fitness of the parent.

Parental investment can be in terms of resources used in the production of gametes, resources used to provide an environment that is suitable for fertilisation or resources used in parental care – the protection and nurture of any offspring produced. Organisms may invest in offspring that have already been produced or offspring that may be produced in the future.

COMPARISON OF SPERM AND EGG PRODUCTION

Parental investment begins with the production of gametes. There is an inequality in the investment required in the production of individual eggs and sperm cells. There is a selection pressure favouring larger eggs, as these will contain an energy store for new offspring. At the same time the sperm have to be fast moving in order to reach an egg rapidly, so there is a selection pressure for them to contain no excess mass.

In humans, females have about 2 million egg follicles, but invest only in about 450 of these; just 1 or 2 are ovulated at any one time. In contrast, males produce about 500 billion mature sperm in a lifetime, of which around 500 million can be ejaculated at one time.

SEX AND SESSILE ORGANISMS

Sessile organisms are those that are fixed in one place. In terrestrial environments, organisms such as plants, lichens and fungi are sessile. In marine environments, the algae and many marine animals are sessile. Sessile organisms often spread by asexual methods, such as budding to form clonal colonies.

For sessile organisms, there is a **problem of sex** that relates to the inability to move and find another organism with whom to exchange gametes. **Solutions** include: being self-fertile, such as in some species of plant that are capable of self-pollination; synchronised spawning, such as in coral, so that gametes are released into the environment at the same time to maximise chances of fertilisation; use of other organisms to carry gametes, such as in insect-pollinated plants.

Comparison of fertilisation strategies.

Type of fertilisation	Description	Costs	Benefits	Note
External	Gametes are released directly into the environment.	Chances of success are lower, so more gametes are required. Little control over choice of mate. Relies on suitable environmental factors.	No increased exposure to predation or parasites.	Many mobile organisms that use external fertilisation only release gametes when in close contact with a potential mate.
Internal	Gametes are released into the reproductive tract of the female.	Finding a mate increases risk of predation. Interacting with conspecifics has increased risk, including those of parasitism.	Fewer gametes need to be released as chances of success are higher. Increased potential for mate choice.	A general trend is that more sperm is produced in species where there is more competition for fertilisation.

VIDEO LINK

Watch synchronised spawning of a mobile externally fertilised species, the horseshoe crab at www.brightredbooks.net

The African elephant is a classic example of a K-selected species. The gestation period is two years and there is an extended period of parental care, as shown here.

R-SELECTED AND K-SELECTED ORGANISMS

This simplistic classification relates to the trade-off between quality and quantity of offspring. Those that are said to be **r-selected** tend to be small species that produce many offspring, each of which has a low chance of survival. This is a successful evolutionary strategy for opportunistic species in unstable environments. **K-selected** species tend to be longer-lived and are more likely to be subject to strong intra-specific competition for resources in stable environments. These species show large parental investment in a much smaller number of offspring.

REPRODUCTIVE STRATEGIES

Various reproductive strategies have evolved, ranging from polygamy to monogamy.

In **polygamy** each individual may mate with several others in the same reproductive season. Lekking (see page 77) species are a good example of polygamy.

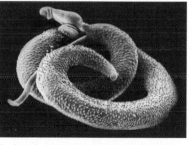

In **monogamy**, each individual only shares gametes with one other individual. DNA fingerprinting studies have revealed that many species that were thought to be monogamous actually have a proportion of extra-pair copulations. The *Schistosoma* parasite is extremely monogamous – the female of the pair lives within a canal-like structure in the body of the male.

The female of the monogamous *Schistosoma mansoni* parasite lives in a groove in the body of the male. The male is larger with a roughened surface. The female is narrower and has a smooth surface.

ONLINE TEST

Test your knowledge of this topic at www.brightredbooks.net

THINGS TO DO AND THINK ABOUT

The bee orchid is a sessile organism that manipulates a drone bee (male) into moving its pollen from flower to flower to achieve pollination. The flower (pictured here) mimics the sign stimulus of the queen bee and the drones are duped into mating with them. Mating with a series of bee-orchid flowers is of no benefit to the drone, but is of great benefit to the plant. More normally, the co-evolution of pollination is mutualistic and the insects are 'paid' by plants with nectar for their pollination duties.

COURTSHIP

COURTSHIP AND FIXED-ACTION PATTERNS

Courtship involves behaviours and characteristics associated with attracting a mate. DNA sequences that result in higher chances of successful breeding are favoured by sexual selection. The phenotypes associated with these sequences are not the same ones that would be selected by natural selection alone – they may not improve an individual's chances of survival. However, an increased chance of successful reproduction can lead to an overall increase in evolutionary fitness. As a result, sexual selection results in the evolution of elaborate and intricate courtship displays and sexual dimorphism.

Many species show ritualised courtship behaviours in which a specific **sign stimulus** from one individual produces a specific **fixed-action pattern response** in the other. This type of response is instinctive and thought to vary little between members of the same species. For example, in the courtship of stickleback, the swollen belly of a female fish acts a sign stimulus to for the male to commence his zig-zag courtship dance. If uninterrupted, a series of sign stimuli and fixed-action pattern responses lead the stickleback pair through the complete courtship to the successful fertilisation of eggs (see also page 59).

IMPRINTING AND COURTSHIP

Mate choice can be influenced by sexual imprinting early in life. Imprinting is an **irreversible** developmental process that occurs during a **critical time period** in young animals. Imprinting increases fitness through natural selection, as forming an attachment to a parent that is providing care increases survival chances. In addition, imprinting can **influence mate choice** later in life as offspring may select mates that resemble parental phenotypes.

The herring gull (left) and lesser black-backed gull (right) are two distinct species that are commonly seen in school grounds. Very rarely, these species interbreed to produce hybrid offspring (centre). The incidence of mixed pairs has been shown to be greatly increased among birds raised by parents of the wrong species as a result of cross-fostering experiments.

The eider is a sexually dimorphic species that is found around the coasts of Scotland. (a) The male has conspicuous plumage, as well as a noisy courtship display. (b) The female has cryptic camouflage which helps protect the eggs and young during parental care. The young are active soon after hatching and imprint on the female.

SEXUAL DIMORPHISM

Sexual selection tends to result in distinct differences between the two genders. This sexual dimorphism is defined as **differences in characteristics** between the two sexes of the same species – other than differences in the sex organs. Typically differences are in terms of size and colouration. In most cases of sexual dimorphism, the male is either larger, more heavily armoured, or more conspicuous in appearance or behaviour. Females are generally inconspicuous; males have more conspicuous markings, structures and behaviours.

REVERSED SEXUAL DIMORPHISM

Some species of birds have reversed sexual dimorphism in which the female is larger or more brightly coloured. In many species of birds of prey, the female is considerably larger than the male – the smaller male is better adapted for hunting small prey that

contd

are suitable for feeding his young nestlings during the breeding season. In a couple of species of Arctic wading birds, the females are brighter than the males; the males, as well as the females, incubate clutches of eggs and raise broods of chicks – an adaptation that increases the reproductive fitness of the individuals during the short breeding season.

MALE–MALE RIVALRY

In many sexually dimorphic species, success in competitive male–male rivalry can increase the access of males to females. The successful males will be those that are stronger or have **greater weaponry**, and these are fitness characteristics that will be advantageous to offspring. Alternatively, some males are successful by acting as **sneakers**. Sneakers can superficially resemble the female in appearance in order to avoid male–male rivalry.

FEMALE CHOICE

In other species, female choice drives the evolution of conspicuous markings, structures or behaviours in males.

The male displays tend to reveal 'honest signals' – characteristics that allow potential mates to assess the genetic quality of the males. Good phenotypic quality is an indicator of 'good genes' and low parasite burden.

LEKKING

Lekking involves **communal displaying** by groups of males. Females visit the lek area to assess males and select a suitable mate. In lekking species, there can be **alternative successful strategies** for males: dominant males tend to be larger and win in the aggressive encounters between rival males; satellite males attempt to intercept females at the periphery of the lek area.

The black grouse is a lekking species found in Scotland. The display reveals the physical fitness, genetic quality and parasite burden of the male (right) as the female assesses the redness of his eye wattle, the cleanliness of his white undertail feathers, the symmetry of the plumage and the intensity of his leaping and calling.

ONLINE

To sneak or not to sneak in giant cuttlefish, www.brightredbooks.net

(a) The male giraffe is considerably taller than the female and has much bigger horns. The male giraffes exhibit male–male rivalry and fight using their heads and necks for access to females (b) Two well-matched male impala resort to physical combat.

ONLINE

Watch the incredible display of birds of paradise at www.brightredbooks.net

THINGS TO DO AND THINK ABOUT

In one lekking species, the ruff, there are three distinct types of male. **Territorial males** are larger than females and have a dark ruff of feathers around the neck. Territorial males fight each other for the best positions in the lek. **Satellite males** are slightly smaller, less intensely coloured and are tolerated at low density by territorial males as having a larger number of males at a lek is advantageous (it attracts more females). **Sneaker males** are the same size and plumage as females. Genetic studies have shown that each type of male is the result of a 4 million year old chromosomal inversion mutation that affects 90 genes, including those for pigmentation and testosterone metabolism. Crossing over cannot occur at inversions in heterozygous individuals, so the 90 genes are passed on together as a single 'supergene'.

ONLINE

Check out this mind-boggling lekking display of the sage grouse at www.brightredbooks.net

ONLINE TEST

Test your knowledge of courtship at www.brightredbooks.net

THE PARASITE NICHE

VIDEO LINK

Watch about resource partitioning at www. brightredbooks.net

ECOLOGICAL NICHE

The concept of the **ecological niche** attempts to summarise all those factors that influence the distribution of an individual species. At its simplest level, niche was originally defined as the role an organism plays in the living community – such as 'woodland herbivore' or 'grassland carnivore'. Clearly no herbivore could survive in all types of woodland, so a more sophisticated definition of niche is required.

The difference between the fundamental niche and realised niche of a species.

The concept of the ecological niche must include many factors (or dimensions). The dimensions that have to be considered include **abiotic factors** (e.g. temperature, humidity, soil pH, salinity, light intensity, pollution, gas concentrations, minerals) as well as **biotic interactions** within the community (e.g. predator–prey interactions, competition, parasitism). The ecological niche is, therefore, best defined as the **multidimensional summary of tolerances and requirements of a species**.

Fundamental and realised niches

- The **fundamental niche** of a species is the set of resources that it is capable of using in the absence of any interspecific competition (competition with other species).

- The **realised niche** is the set of resources that the species *actually* uses in response to the presence of interspecific competition.

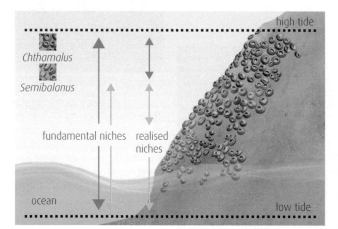

Shoreline zonation – the niche at its most visual.

The fundamental and realised niches of two species of barnacles can be seen in terms of their positions on the intertidal shoreline. The fundamental niche of *Chthamalus* is very wide and extends from the low- to high-tide line. The fundamental niche of *Semibalanus* is only on the lower part of the shore. When the two species are in competition, *Chthamalus* becomes restricted to a realised niche on the upper shore.

Competitive-exclusion principle

DON'T FORGET

The fundamental niche is the *theoretical* niche, whereas the realised niche is the *reality*.

In some situations, where two species are competing, the realised niches of the two species may be very similar. One species will lose out more in the competition and so its population will decline, leading to **local extinction**. This observation is known as the **competitive-exclusion principle**. For example, grey squirrels, which were introduced into the UK, are able to eat seeds from deciduous trees before they are fully ripe and so are able to out-compete red squirrels in these habitats.

Resource partitioning

Where two species living in a habitat have similar requirements, it may be that the realised niches are sufficiently different for them to co-exist. In this case, **resource partitioning** can occur whereby the two species exploit different components of the resource, thus reducing potential competition. The evolution of different beak lengths and foraging behaviours in waders has allowed various species to exploit food of different types and at different depths in their shared habitat.

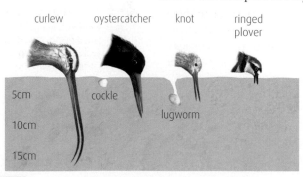

The beaks and foraging depths of wading birds.

THE NICHE OF PARASITES

Symbiosis ('together living') refers to intimate associations between individual organisms of different species. Parasites are symbionts that gain nutrients at the expense of the host. The parasite uses the host's resources for growth and reproduction; the host, as well as losing resources, also incurs further costs in defending its tissues from parasitic attack. **Ectoparasites** live and feed on the surface of their host, e.g. ticks, lice, fleas. **Endoparasites** live within the host, e.g. tapeworm, *Plasmodium* (causes malaria), rhinovirus (causes common cold).

Human crab louse – an ectoparasite which is highly specific to the pubic hair of humans.

In a predator–prey relationship, the predator and the prey have similar reproductive potential, e.g. cats and mice can both produce tens of offspring each year. In a parasitic relationship, the reproductive potential of the parasite is much greater than that of the host, e.g. a cat flea can produce thousands of offspring in a year.

Co-evolution and the parasite niche

According to the Red Queen Hypothesis, parasite adaptations have been selected in response to the particular adaptations that have evolved in the host species. Because of this, parasites tend to have a **narrow niche**: this is true both in terms of being **host-specific** and also very specific in their ways of exploiting that host. For example, there are three ectoparasitic species of louse which are only found on humans, and each louse species has a restricted area of the body on which they feed – an example of resource partitioning.

Many of a parasite's needs are provided by the host, so this means that parasites have become **degenerate** – they lack structures and organs that are found in other organisms. For example, the tapeworm is an endoparasite that lives in the small intestine of its host and is surrounded by digested food, so it has no need for a digestive system. The production of unused tissue in an organism is a waste of resources, so evolution has favoured the loss of these tissues.

Parasite-lifecycle definitions

Parasitic lifecycles can involve more than one species. The **definitive host** is where the parasite reaches sexual maturity and so carries out sexual reproduction. An **intermediate host** is one in which developmental stages happen to complete the parasite's lifecycle. For example, the pork tapeworm reproduces sexually in the gut cavity of humans (the definitive host). The tapeworm sheds eggs into the gut cavity and these eggs pass out in faeces. The eggs are eaten by pigs (the intermediate host) where they hatch and develop as slowly in sacs called cysts in the muscle tissue before being consumed by humans.

Parasite lifecycles may also have species that play an **active part** in the transmission of the parasite. These are called **vectors** and they may also be hosts. For example, the mosquito is the vector for the malarial parasite, and it is also the definitive host. But vectors are not always hosts; aphids are vectors for plant viruses, but the viruses have no lifecycle stages in the aphids.

ONLINE

Find out more about the three types of human louse at www.brightredbooks.net

Lifecycles of four species of human tapeworm. One species relies on direct re-infection, while three species require intermediate hosts. None of the lifecycles have vectors.

DON'T FORGET

The terms *definitive host* and *intermediate host* describe different aspects of a parasite lifecycle, and either of these host types could also be a *vector*.

VIDEO LINK

Check out the cartoon-style explanation of host types at www.brightredbooks.net

ONLINE TEST

Head to www.brightredbooks.net to test yourself on the parasite niche.

THINGS TO DO AND THINK ABOUT

Many parasites carry out asexual reproduction in the intermediate hosts; why would this be a benefit to the parasite?

The population of the parasite is rapidly built up by asexual reproduction so there are many more parasites present to increase the chance of dispersing to a new definitive host.

TRANSMISSION AND VIRULENCE

The spread of a parasite to a new host is called **transmission**. The harm that a parasite causes to a host species is called **virulence**. This reduces the host's evolutionary fitness because the parasite is redirecting the host's resources to parasite reproduction.

TRANSMISSION DETERMINES THE VIRULENCE OF A PARASITE

The parasite's evolutionary fitness is determined by how successful it is in transmitting offspring to new hosts. Selection will favour a combination of virulence and transmission that can maximise the spread of large numbers of the parasite's offspring.

Evolutionarily stable strategies

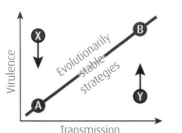

The link between transmission and virulence

Strategy A (low transmission, low virulence) is shown by the rhinovirus which causes the common cold. This parasite relies on direct transmission as it can only survive for a maximum of 3 hours in the environment. High virulence would hamper transmission – if you are so sick with a cold that you cannot leave your house, then the parasite is less likely to infect new hosts.

Strategy B (high transmission, high virulence) is shown by cholera and malaria. These both multiply in the host and are highly virulent (cause great damage), but efficient transmission ensures their spread.

Unsuccessful strategies

Strategy X and Y are not evolutionarily stable. A parasite with strategy X has high reproduction in the host, causing high virulence, but is not transmitted sufficiently well to find new hosts to maintain the parasite population. In this case, selection will favour a strain with a lower virulence, so moving towards strategy A.

Any parasite strain which shows strategy Y will be out-competed by a more virulent strain of the parasite, so moving towards strategy B.

DON'T FORGET

Higher transmission allows the evolution of higher virulence.

FACTORS THAT INCREASE TRANSMISSION RATES

If a population of hosts is **overcrowded** and living at **high density** in a habitat, then parasites are able to pass more easily to a new host. This is important for parasites that are transmitted by direct contact and so tend to have a low transmission rate. For example, the human head louse is more rapidly spread when people have their heads close together, such as in schools...

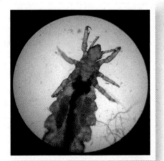

Human head louse, showing human blood in the digestive system and hook-like claws to hold on to fine human hair.

The use of **vectors** has evolved in some parasite lifecycles. This increases the transmission rate as it allows the parasite to spread, even when the infected host is incapacitated. For example, the mosquito can spread the *Plasmodium* parasite from people who are immobilised due to a malarial fever. Some parasites have **waterborne dispersal stages** which take advantage of the liquid medium to disperse and transmit to new hosts. For example, cholera victims are quickly immobilised by the disease, but their waste transmits the bacterium back into the water supply, where it can be taken in by new hosts.

Host behaviour is often exploited and modified by parasites

Some parasites **exploit the natural behaviours** of the hosts to maximise transmission. For example, *Schistosoma* exploits the behaviour of mammals which wade into lakes. The parasite is released from an aquatic snail (intermediate host), so it has a waterborne dispersal stage which can burrow into the skin of the legs of new hosts in the water.

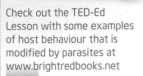

VIDEO LINK

Check out the TED-Ed Lesson with some examples of host behaviour that is modified by parasites at www.brightredbooks.net

contd

The **host's behaviour can be modified** by infection so that the parasite is passed on more easily. In this case, the host behaviour becomes part of the **extended phenotype** of the parasite and there are five key categories of behaviour modification.

Category of behaviour modified	Host and parasite	Description
Foraging	Mosquito and *Plasmodium*	Mosquito with mature parasites is more likely to feed on blood from more than one person
Movement	Frogs and flatworm	Frog develops additional back legs, so moves slower and is more easily caught by the next host (predatory bird)
Sexual behaviour	Mayfly and nematode	Parasite has to return to water to find its next host (fish) so mayfly females go to lay eggs even if no eggs present; males behave like females and attempt to lay eggs
Habitat choice	Ant and flatworm	Ant climbs to top of blade of grass at night, instead of going to nest, so is eaten by its next host (deer or sheep)
Anti-predator behaviour	Rats and Toxoplasma	Rat seeks out the smell of cat urine, so is eaten by cat and parasite is ingested by new host

FACTORS THAT INCREASE VIRULENCE

A parasite with a high transmission rate will experience selection pressure for increased virulence, increasing **parasite growth** and **reproduction** in the host. There are three major ways that endoparasites achieve this. They:

1. suppress the **host immune system** so that parasites survive and reproduce in the host without being attacked by the host's defences

2. modify the **size of the host** so that it grows much larger and can support the asexual reproduction of more parasites, e.g. infected mud snails in Asia grow about 25% larger

3. reduce the host's **reproductive rate** so more of the host's energy can be directed to parasite reproduction, e.g. *Sacculina* is a parasitic barnacle that destroys the gonads of its crab host.

Sexual and asexual phases of lifecycles

The sexual phase of the parasite lifecycle happens in the definitive host. The advantage of the sexual phase is that it produces genetic variation and so allows the **rapid evolution** needed in the Red Queen's race between parasite and host.

The advantage of the asexual phase is that it allows the **rapid build-up** of a parasite population for dispersal into the next host. For example, *Schistosoma* reproduces asexually in the snail before being spread into the water of the lake.

The distribution of parasites is not uniform across hosts

In the graph, most of the host population carry no ticks and a few individuals carry most of the parasite load. This may be because the parasite has a low transmission rate within the population. Perhaps the affected individuals are living in a higher density in the habitat, or show behaviours that increase their chance of being infected.

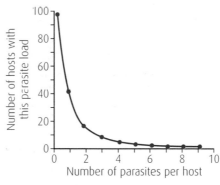

Distribution of ticks (*Ixodes tringuliceps*) in a population of mice (*Apodemus sylvaticus*).

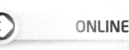 **DON'T FORGET**

You don't need to learn the specific examples of these extended phenotypes, just the five key categories.

ONLINE

Learn more about mud-snail gigantism at www.brightredbooks.net

ONLINE

Read the *Sacculina* horror story at www.brightredbooks.net

 **THINGS TO DO AND THINK ABOUT**

Public-health measures often try to reduce the transmission of parasites. If these measures are successful, what long-term effect will this have on the virulence of the parasite? This effect was observed in cholera outbreaks in South America: when cholera invaded countries with poor water sanitation, the strains evolved to be more virulent, while strains that invaded areas with better sanitation evolved to be less harmful.

ONLINE TEST

Test yourself on transmission and virulence at www.brightredbooks.net

PARASITIC LIFECYCLES

COMMON PARASITIC GROUPS

The diagram shows the main six types of parasites that you need to know. However, parasitism is such a successful strategy that many other groups of organisms have parasitic members, such as: annelids (e.g. leech), fungi (e.g. potato blight), angiosperms (e.g. bird's nest orchid), mammals (e.g. vampire bat) and fish (e.g. lamprey).

Protists e.g. Plasmodium, gut amoeba

Arthropods e.g. ticks, lice

Platyhelminths e.g. Schistosoma, tapeworm

Nematodes e.g. threadworm, elephantiasis

Viruses e.g. influenza, HIV

Bacteria e.g. tuberculosis, syphilis

LIFECYCLES AND HOSTS

Some parasites have only one host species

Some parasites that use direct contact as their transmission mechanism can complete their lifecycle within one host species. Many of these are ectoparasites or endoparasites of the main body cavities (such as the gut or lungs). Since direct contact is a relatively inefficient mode of transmission, these parasites tend to have low virulence.

- Ectoparasitic arthropods (e.g. human head lice) go from egg through to full reproductive maturity on a single host. This is why the removal of adult head lice and eggs demands repeated combing over a number of weeks. Transmission is achieved by adults moving to new hosts in close proximity.

- Endoparasitic amoebas (e.g. *Entamoeba histolytica*) complete their lifecycle from start to finish in the large intestine of their human hosts. Transmission is achieved using cysts in faeces; these cysts are taken up by the next host in contaminated food or water.

- Bacteria (e.g. *Mycobacterium tuberculosis* which causes tuberculosis) and viruses (e.g. influenza virus) can repeat their lifecycle many times in one host. When the host breathes out, the parasites are transmitted in droplets in the air, so the bacterium or virus comes into direct contact with a new host.

Many parasites need more than one host species

Some parasites have to spend part of their lifecycle in an intermediate host species to complete the next stage of their development, so that they are better able to be transmitted to a new definitive host. Because this improves the transmission efficiency, these parasites have a higher virulence.

Some ectoparasite and some endoparasite species have increased their transmission efficiency by having a lifecycle stage in an intermediate host. Transmission is achieved when the definitive host **consumes the intermediate host**. For example, the eggs of the pork tapeworm are passed out from a person in the faeces and the eggs are eaten by pigs (the intermediate host). The eggs hatch and develop as cysts in the muscle tissue, before being consumed by a human.

contd

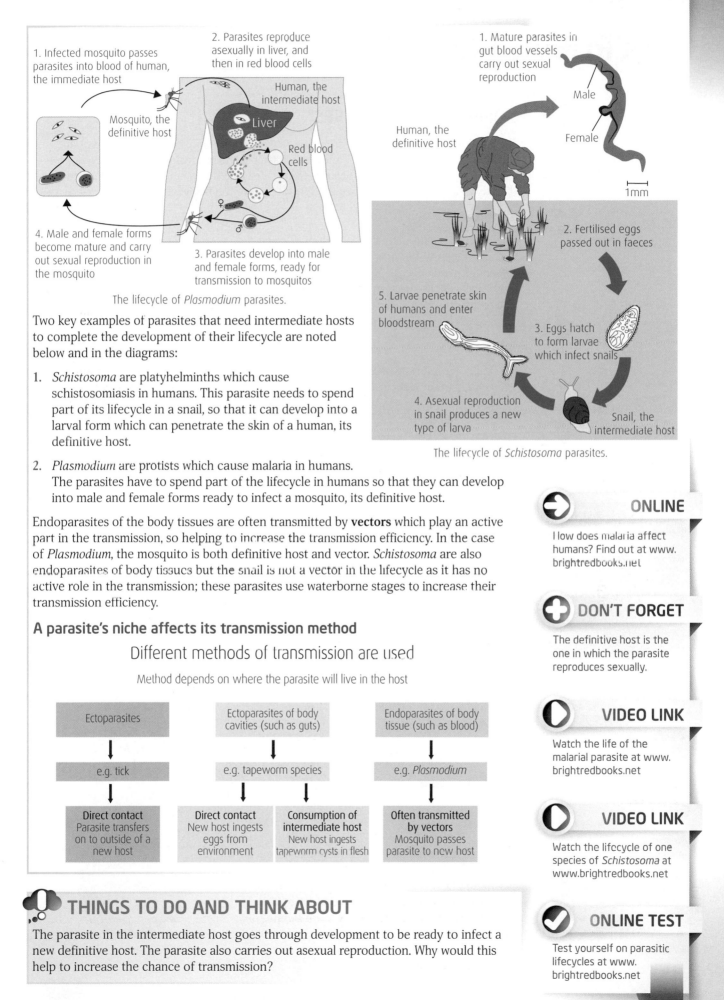

The lifecycle of *Plasmodium* parasites.

The lifecycle of *Schistosoma* parasites.

Two key examples of parasites that need intermediate hosts to complete the development of their lifecycle are noted below and in the diagrams:

1. *Schistosoma* are platyhelminths which cause schistosomiasis in humans. This parasite needs to spend part of its lifecycle in a snail, so that it can develop into a larval form which can penetrate the skin of a human, its definitive host.

2. *Plasmodium* are protists which cause malaria in humans. The parasites have to spend part of the lifecycle in humans so that they can develop into male and female forms ready to infect a mosquito, its definitive host.

Endoparasites of the body tissues are often transmitted by **vectors** which play an active part in the transmission, so helping to increase the transmission efficiency. In the case of *Plasmodium*, the mosquito is both definitive host and vector. *Schistosoma* are also endoparasites of body tissues but the snail is not a vector in the lifecycle as it has no active role in the transmission; these parasites use waterborne stages to increase their transmission efficiency.

A parasite's niche affects its transmission method

Different methods of transmission are used

Method depends on where the parasite will live in the host

Ectoparasites	Ectoparasites of body cavities (such as guts)		Endoparasites of body tissue (such as blood)
↓	↓		↓
e.g. tick	e.g. tapeworm species		e.g. *Plasmodium*
↓	↓	↓	↓
Direct contact Parasite transfers on to outside of a new host	**Direct contact** New host ingests eggs from environment	**Consumption of intermediate host** New host ingests tapeworm cysts in flesh	**Often transmitted by vectors** Mosquito passes parasite to new host

THINGS TO DO AND THINK ABOUT

The parasite in the intermediate host goes through development to be ready to infect a new definitive host. The parasite also carries out asexual reproduction. Why would this help to increase the chance of transmission?

ONLINE

How does malaria affect humans? Find out at www.brightredbooks.net

DON'T FORGET

The definitive host is the one in which the parasite reproduces sexually.

VIDEO LINK

Watch the life of the malarial parasite at www.brightredbooks.net

VIDEO LINK

Watch the lifecycle of one species of *Schistosoma* at www.brightredbooks.net

ONLINE TEST

Test yourself on parasitic lifecycles at www.brightredbooks.net

MICROBIAL PARASITES

DON'T FORGET

Most microbial parasites have only one host, so use direct contact for transmission.

ONLINE

Read more about Lyme's disease at www. brightredbooks.net

ONLINE

Read more about tuberculosis at www. brightredbooks.net

ONLINE

Read more about influenza at www.brightredbooks.net

PREVENT DISEASE

CARELESS
SPITTING, COUGHING, SNEEZING,
SPREAD INFLUENZA
and TUBERCULOSIS

American public health poster from 1918

ONLINE

Read more about HIV/AIDS at www.brightredbooks.net

DON'T FORGET

The genetic material of viruses can be DNA or RNA.

SOME HUMAN DISEASES

Tuberculosis, influenza and HIV are caused by microbial parasites. Most microbes are transmitted by direct contact, so they only have **one host species** in their lifecycle. One exception is the bacterium that causes Lyme's disease; this is cannot be transmitted directly to a new host, it is transmitted by bites from the tick vector.

Tuberculosis

This disease is caused by the bacterium, *Mycobacterium tuberculosis*. It affects many parts of the body, but mainly the lungs. It is transmitted by inhaling tiny droplets from the coughs and sneezes of an infected person. The immune system of a healthy person can usually kill the bacterium, or at least stop it spreading. The disease causes about 1.5 million deaths every year; most of these are in areas where people have lower health status such as sub-Saharan Africa.

Influenza

The 'flu is caused by a virus which attacks joints, muscles and the respiratory tract. It is transmitted in droplets from coughs and sneezes. The symptoms last for about a week and most people recover, though about 500 000 people die each year from 'flu. The most deadly human conflict was World War I which killed 17 million people in four years; a flu epidemic in the two years just after the end of the war is estimated to have killed between 20 and 100 million people worldwide.

Human immunodeficiency virus (HIV)

This virus is transmitted in bodily fluids such as blood, semen, vaginal fluid or breast milk. In the UK, it is most commonly transmitted by having sex without a condom. The virus attacks T-helper lymphocytes of the immune system (see page 87) and when too many of these cells are infected, the body can no longer fight any infections. AIDS is the final stage of HIV infection, when the body can no longer fight everyday infections, which become life-threatening. Early diagnosis and treatment means that most people with HIV will not go on to develop AIDS.

THE 'LIFE' OF A VIRUS

Viruses are infectious agents that can only replicate inside a host cell. They don't quite meet most biologists' definition of 'living'. The viruses can't carry out any of the normal functions for life except for one – reproduction, and they can only do this in the cell of a living organism.

Virus structure

For reproduction to occur, viruses have to contain genetic information stored in a **nucleic acid** (DNA or RNA). This nucleic acid is packaged inside a **protective protein coat** (the capsid) and some viruses also have an outer envelope made from part of the host-cell membrane. The outer surface of a virus has proteins that are coded for by the viral genes. These proteins are called **antigens**, because a cell from the host organism is able to detect them as foreign and can, therefore, initiate an immune response.

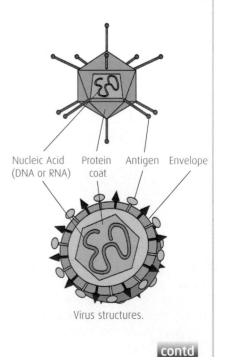

Nucleic Acid (DNA or RNA)   Protein coat   Antigen   Envelope

Virus structures.

contd

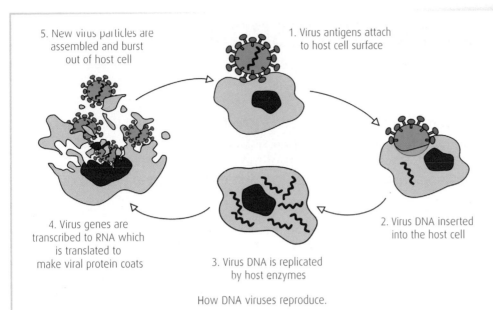

5. New virus particles are assembled and burst out of host cell

1. Virus antigens attach to host cell surface

4. Virus genes are transcribed to RNA which is translated to make viral protein coats

2. Virus DNA inserted into the host cell

3. Virus DNA is replicated by host enzymes

How DNA viruses reproduce.

Virus reproduction

DNA viruses have their genetic material in the form of DNA and these include smallpox, herpes, and chickenpox viruses. These viruses reproduce using the steps shown in the diagram.

RNA viruses (such as influenza, Ebola, hepatitis C, polio, rabies and measles) have an RNA genome. These viruses use the same process of reproduction as the DNA viruses except, at Step 3, the viral RNA genome is replicated directly using an enzyme from the virus. Viral DNA is never made.

RNA retroviruses, such as HIV, use a slightly more complex process for reproduction, though it is really just Step 3 that is different. These viruses use the enzyme **reverse transcriptase** to form a DNA copy of the virus genome. This DNA is then integrated into the genome of the host cell. The virus genome can then be replicated as part of the normal cell cycle, so the virus genome will always be in the host cells. When the virus is reproducing, the DNA copy of the virus genome will be transcribed to RNA, which is then translated to make new viral proteins. These proteins and the RNA form new virus particles and these then burst out of the host cell.

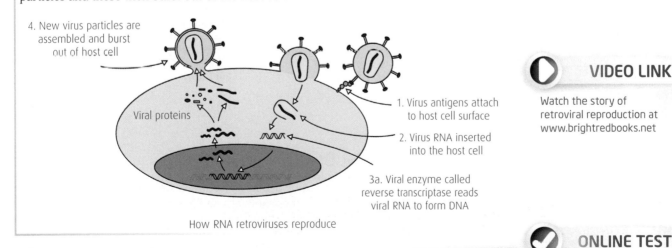

4. New virus particles are assembled and burst out of host cell

Viral proteins

1. Virus antigens attach to host cell surface

2. Virus RNA inserted into the host cell

3a. Viral enzyme called reverse transcriptase reads viral RNA to form DNA

How RNA retroviruses reproduce

THINGS TO DO AND THINK ABOUT

The treatment of HIV infection has become very effective at preventing the development of AIDS in the patient. The treatment uses a combination of three or more antiretroviral drugs to stop the virus replicating. The combination of drugs is needed because a single antiretroviral drug is a simple selection pressure, so the virus can quickly evolve and become resistant to that drug.

MAMMALIAN IMMUNE RESPONSE TO PARASITES

Mammals have evolved a layered defence against parasites. The first and second line of defences are **non-specific** and work against of a wide range of parasites. The third line is a cellular response that is **specific**, because it attacks the antigens of the parasite. However, the battle may not be won by the host as parasites have evolved their own strategies to evade host defences.

NON-SPECIFIC DEFENCES

First-line defences

These are **physical barriers** and **chemical secretions** that work together to prevent many parasites from entering the body fluids.

Physical barriers: The skin acts a tough barrier against entry by many parasites; it is also dry, so many microbial parasites die on the surface. Nasal hairs trap many microbial parasites and prevent them from contacting the delicate tissues in the lungs.

Chemical secretions: Mucus in the lungs traps microbial parasites. The mucus is continuously moved away from delicate alveolar tissue by the action of cilia. Nasal mucus and earwax trap microbial parasites and these either seep away or dry up and flake off. Tears have antibodies and lysozyme enzyme to defend against microbial parasites. The skin and the stomach have acidic secretions that kill most microorganisms.

Second-line defences

Second-line defences happen as a response after a parasite has entered the body fluids. They are produced quite rapidly without the need to determine the invader's identity, i.e. they are non-specific.

Some of the phagocytes remain active. After they have engulfed and digested the foreign object, they present **fragments of the foreign antigens** on their surface. This will be important as part of the immune surveillance in the next stage of defence.

Inflammatory response	Phagocytes	Natural killer cells
Some injured or infected cells can release chemicals including **histamine**.	These are white blood cells which are able to migrate out from the blood into the tissue fluid which surrounds cells. The phagocytes can engulf bacteria, viruses and dust particles.	These are also a type of white blood cell which can migrate into the tissue fluid. They detect the **abnormal cell-surface proteins** found on virus-infected cells and on cancerous cells.
This causes the local blood vessels to dilate, so increasing the blood flow to the area, making it become red and warm.		
The permeability of the blood vessels also increases, so fluid moves into the tissue, which swells up; this puts pressure on nerve endings leading to pain in the area.	The phagocytes check the surfaces of cells and particles for 'self' antigens. If these are not found, then the phagocyte **engulfs and digests** the foreign object.	The natural killer cells do not engulf the stricken cell but, instead, attach to it and release chemicals into it. These chemicals induce the cell to kill itself, a process called **apoptosis**.
This swelling also stimulates **phagocytes** to migrate to the area.	Many phagocytes die at the site of their action and form the pus that is found at an infected wound.	Phagocytes engulf and digest the resulting cell debris.

SPECIFIC CELLULAR DEFENCES

The third line of defences is triggered by the antigens on the surface of the parasite. The responses at this stage are specific to these antigens and involve lymphocytes.

DON'T FORGET

Specific defences are produced in response to the presence of a specific antigen.

Lymphocytes and immune surveillance

Lymphocytes are white blood cells which are found mainly in the lymph glands. Lymphocytes each present only one type of **antigen-receptor protein** on their surface and these can bind to just one specific antigen. Although each of us can produce between 1 million and 10 million different antigen receptors, each lymphocyte can make only one type of receptor. This is because each lymphocyte is part of a **clone**, a group of about 1000 identical cells made from a common ancestral cell, which was committed to produce a single type of antigen receptor.

contd

Tissue fluid is drained from the around the body's cells by the lymphatic system. The lymph fluid passes through the lymph glands, where the lymphocytes carry out **immune surveillance**, checking for their specific antigens using the receptor proteins on their surface. During this surveillance, phagocytes in the lymph fluid are also checked for their presented antigen fragments.

VIDEO LINK

An animated summary of the immune system can be found at www.brightredbooks.net

Clonal selection and the action of lymphocytes

The specific lymphocyte clones undergo **clonal selection** when their antigen-receptor proteins bind to their specific antigen. This activates the lymphocytes, making them ready to divide rapidly to produce many more clones, each with the same specific antigen receptor.

Clonal selection happens in two different ways. **B lymphocytes** are selected when their antigen-receptor proteins bind directly to their specific antigen. **T lymphocytes** can only be activated if phagocytes present the correct antigen fragments to bind with their receptor proteins.

One class of T lymphocyte, the **helper T lymphocytes**, immediately begin to divide rapidly and produce many clones of themselves. These helper T lymphocytes then **target the immune-response cells**, which means that they stimulate the B lymphocytes and cytotoxic T lymphocytes to divide rapidly.

Cytotoxic T lymphocyte

B lymphocyte

The B lymphocytes and cytotoxic T lymphocytes have quite different roles.

1. **B lymphocytes** produce **antibodies**, proteins with the same binding sites as the receptor proteins. These antibodies bind to the antigens on the parasite. This neutralises the parasite and makes it easier for phagocytes to find and engulf the parasite.

2. **Cytotoxic T lymphocytes** destroy infected cells by inducing **apoptosis**. (Although this is a similar function to that of the natural killer cells, the cytotoxic T lymphocytes are much more effective but can only attack a cell if it has the specific antigen.)

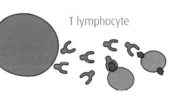

T lymphocyte

Immunological memory

Once the parasite infection has been defeated, some members of the T lymphocyte and B lymphocyte clones are kept as an **immunological memory**. The **long-term survival** of a larger population of these lymphocyte clones means that they produce a quicker and larger response if the specific antigen is detected again. This is the basis of immunisations, whereby an antigen is presented to the immune system using a harmless version of the parasite.

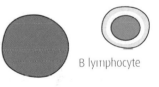

B lymphocyte

SOME PARASITES CAN RESIST THE IMMUNE SYSTEM

The ongoing Red Queen's race between the parasite and host means that some endoparasites have evolved ways to evade the immune system.

1. Mimic host antigens – hide from the immune system	2. Antigenic variation – be a moving target for the immune system	3. Modify host's immune response – reduce the chances of being destroyed
Schistosoma produces surface antigens that mimic those of the human host. This helps it to evade detection by the immune system and so there is a reduced antibody response by the host.	*Plasmodium* have hundreds of copies of surface-antigen genes and each new generation expresses a new, randomly selected set of these antigen genes. This allows the parasite to evolve quicker than the host immune system can respond, so the parasite is always one step ahead.	*Schistosoma* has mechanisms to degrade the specific antibodies that have been produced in defence against it. *Plasmodium* can suppress the activity of the specific B and T lymphocytes that have been produced to attack them.

THINGS TO DO AND THINK ABOUT

Nettle stings produce a very quick inflammatory response, because the sting contains histamine which automatically causes the body's inflammation response. The sting also injects acetyl choline and serotonin, which are neurotransmitters that cause the intense pain of a nettle sting.

ONLINE TEST

Test yourself on the mammalian immune response to parasites at www.brightredbooks.net

TREATMENT AND CONTROL OF PARASITES

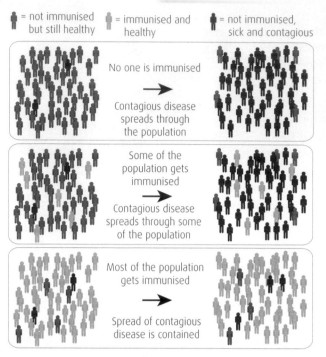

ⓜ = not immunised but still healthy   **ⓜ** = immunised and healthy   **ⓜ** = not immunised, sick and contagious

No one is immunised

→

Contagious disease spreads through the population

Some of the population gets immunised

→

Contagious disease spreads through some of the population

Most of the population gets immunised

→

Spread of contagious disease is contained

Herd immunity.

EPIDEMIOLOGY

Epidemiology is the study of the **outbreak and spread of infectious disease**. The information gained from epidemiological surveys can be used to plan and evaluate strategies to prevent the spread of the parasite in future.

One way to prevent the spread of disease is to use vaccinations to increase the number of resistant hosts in the population; to create **herd immunity**. This occurs when any new infection in a population can be contained because susceptible hosts are too dispersed for the parasite to continue to spread. The **herd-immunity threshold** is the density of resistant hosts required in a population to prevent an epidemic.

Many vaccination programs aim to reach the herd-immunity threshold as this protects any individuals who cannot have the vaccination for health reasons. This is particularly important in less economically developed countries, where poorer nutrition weakens the health of individuals so they are more susceptible to diseases and their complications.

PROBLEMS IN TREATING PARASITIC DISEASES

Vaccination

Vaccination exposes the immune system to a non-infectious source of parasite antigens. This develops immunological memory to these antigens and so protects against infection by the real parasite. This is a relatively straightforward concept, but developing a new vaccine is not so easy in practice.

> Some parasites (e.g. *Schistosoma*) are **difficult to culture** in the laboratory because they need a very narrow range of conditions, as well as specific host signals for their development.

> The parasite may use **rapid antigen change** as a means of evading the immune system (e.g. *Plasmodium*) and this makes **vaccine design very complex;** a vaccine would have to offer many possible antigens.

Other microbial parasites may have rapidly evolving surface antigens (e.g. influenza), so new vaccines have to be developed every year.

One strategy for vaccine development targets the surface proteins needed for fundamental aspects of the parasite's infective action. Including these proteins as the antigens in a vaccine may lead to vaccines with long-term effectiveness; these proteins tend to be very highly conserved due to the importance of their role for the parasite.

Drugs

Drugs that work to kill the parasite in the host could have detrimental effects on the host, due to the similarities between the **host and parasite metabolism**. This was a major problem with early antiparasitic drugs, which were difficult to administer in effective doses that were still safe to the patient. The difficulty for researchers is to find new drug compounds that **only target the parasite**. These problems meant that only 0·1% of all new drugs from 1975 to 1999 were antiparasitics.

VIDEO LINK

Check out the clip about herd immunity at www.brightredbooks.net

ONLINE

Read about the near success of polio vaccination programmes and worries about a lack of herd immunity in Europe at www.brightredbooks.net

DON'T FORGET

You need to know the definition of herd-immunity threshold.

DON'T FORGET

You need to know the two problems associated with developing antiparasitic drugs.

contd

Investment by charities and use of molecular-biology techniques has increased the number of antiparasitic drugs being found, but they are still a tiny minority of medicines under development. Most parasitic diseases are prevalent only in the poorer parts of the world, so there has been little possibility of pharmaceutical companies recouping the huge investment costs of developing a new drug.

A further problem is that drugs provide a selection pressure, leading to the evolution of resistant strains of parasites. New drugs are used in carefully planned combinations so that, even if the parasite could resist one drug, it will be killed by the others.

CONTROLLING PARASITES

Successful parasites are always running in the Red Queen's biological race so, while our medical efforts against them may give us the upper hand in the short term, it is likely that many parasites will continue to evolve countermeasures. In many situations, it may be that the only effective control strategy is to attack the parasite indirectly, by working on ways to reduce the spread of parasites.

Targetting transmission

Many parasites use waterborne stages or vectors to increase their transmission rate, so these are potential targets. If effective, blocking these may have the added benefit that they act against other parasites that use the same transmission routes.

1. **Civil-engineering projects** (such as installing toilet facilities, sewerage systems and water taps from boreholes) all work to **improve sanitation**. These prevent the mixing of drinking water with water that is contaminated by human waste, blocking the transmission route for many different parasites, such as cholera bacteria and *Schistosoma*.

2. Using a **coordinated vector-control strategy** can be highly effective. Measures to reduce the spread of malaria include spraying houses with insecticide to kill mosquitos and using bed nets to prevent access of mosquitos to people at night. These have also reduced the spread of other mosquito-borne diseases like Dengue fever, yellow fever and elephantiasis.

Reducing the transmission rate may be very difficult in situations where parasites have conditions that help them to **spread rapidly**. Challenges arise in parasite control as a result of **overcrowding** (e.g. in refugee camps or shanty towns) and **tropical climates** where the parasites can reproduce rapidly in warm environmental conditions.

Benefits of improving parasite control

Many charities run campaigns that rightly focus on the terrible effects of parasitic infections on individuals. These charities invest in better parasite-control measures and so greatly **reduce child mortality**, but there are wider benefits to the population. The general health of children is greatly improved, since their bodily resources are not being redirected to support the parasite. The children, therefore, have **more resources for growth and development** and, since more children are healthy, this leads to **population-wide improvements** in **child development and intelligence**.

Charity campaigns aim at our individual sensibilities, but the effect of the charities' actions are much deeper and broader.

💭 THINGS TO DO AND THINK ABOUT

The parasite-control double whammy
Parasite virulence is high in a population with poor health. Improving the health of the population through better nutrition, hygiene and sanitation is an effective way to reduce virulence. Better hygiene and sanitation also reduce the rate of transmission – and lower transmission rates force the parasite to evolve lower virulence.

INVESTIGATIVE BIOLOGY

SCIENTIFIC METHOD AND ETHICS

DON'T FORGET

The type of analysis of data to be used should inform the design of the experiment from the outset. Trying to 'bolt on' a statistical test to a poor design is unlikely to improve reliablility.

THE SCIENTIFIC CYCLE

Scientific knowledge is constantly evolving and changing as new data is generated, leading to refinement of our current understanding and forming of new questions. The scientific cycle is a series of steps that biologists use to investigate phenomena in a **testable**, **measurable** and **reproducible** way, with the ultimate aim of explaining and predicting future observations by refining a **hypothesis**.

Construction of a hypothesis

A hypothesis is a potential explanation for an observed event, or a predicted outcome for a proposed experimental procedure. It must be possible to test a hypothesis using the steps in the scientific cycle and it should allow prediction of future events. The **null hypothesis** is an important starting point in the scientific cycle which proposes that there is no link between the independent and dependent variables in an experiment.

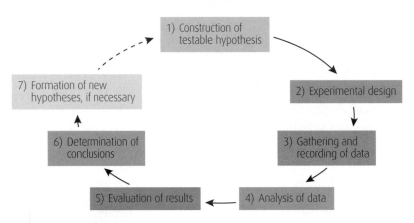

Stages in the scientific cycle.

Experimental design

Design of experiments must provide data to confirm or refute the hypothesis. Care should be taken to ensure experiments are **reproducible**, **reliable**, include both **negative** and **positive control groups** where appropriate, and make sure all important **confounding variables** are kept constant or monitored.

DON'T FORGET

A **hypothesis** and a **theory** are not the same. A hypothesis only becomes a theory when it has been repeatedly tested and confirmed, until the evidence is strong enough to allow the hypothesis to become generally accepted.

Gathering and recording of data

Raw data should be recorded to an appropriate degree of accuracy, which is determined by the scale of measurements being made, limitations of equipment and a balance between cost, time and the number of measurements required for statistical analysis.

Analysis of data

Once gathered, data should be analysed to determine whether conclusions can be formulated. Analysis of experimental data should include consideration of the accuracy and precision of measurements, of the number of replicates and may also include an appropriate statistical test.

Determination of conclusions

After data analysis, it should be decided whether the data supports the hypothesis or if a new hypothesis should be proposed to explain the results. It is important to recognise that failure to find an effect is a **valid finding** if the experiment was well designed.

DON'T FORGET

A **negative control** provides data in the absence of treatment. A **positive control** ensures that the experimental design can detect a positive result when it occurs. **Confounding variables** are variables, other than the independent variable being tested, that may have an effect on the dependent variable being measured.

SCIENTIFIC ETHICS

Ethics can be thought of as the rules for distinguishing between right and wrong. All science should be carried out with a strong ethical basis, given the power of science to impact on society and the environment. Scientific ethics relate to both the internal viewpoint of the scientist and the external rules imposed by society or governments.

Personal ethics of the scientist

Each individual scientist must take responsibility for carrying out their work in an ethical manner. As such, results should be presented in an **unbiased** way by including all data.

contd

Scientists should always acknowledge the contributions of others to their work. Where previous studies are quoted or summarised, the original author must be cited and referenced. A **citation** is an acknowledgement that the idea or findings being discussed are those of another scientist and this appears in the text where the previous work is being described or quoted. A list of **references** for all citations should be included in any scientific writing and must contain enough information to allow readers to find the original publication. Appropriate use of citations and references is essential for avoiding **plagiarism**.

Ethics of animal studies

Many biological studies require the use of animal experimentation, due to the complex and interdependent nature of different cells and systems which cannot always be modelled outside the body. The use of animals for research is controversial and, as such, scientists should be governed by the principles of **Replacement**, **Reduction** and **Refinement**:

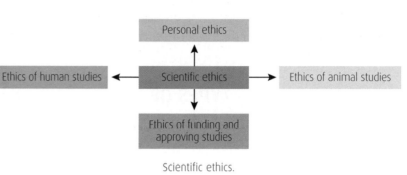

Scientific ethics.

Replacement – use alternatives to animal experimentation when an alternative cellular or model system exists.

Reduction – use the minimum number of animals in a study (without compromising the validity or reliability).

Refinement – techniques should be adjusted to minimise negative impact on the animals.

Ethics of human studies

Any experimentation on humans is governed by four key principles:

1. Participants should be given details of the study to enable them to agree to take part, known as giving their **informed consent,** and must have the opportunity to decline to be involved.

2. Participants have the **right to withdraw data** and their participation at any point during the study.

3. Any data gathered should be anonymous and kept **confidential**.

4. Lastly, studies should aim to ensure that participants come to **no harm** from the study.

Ethics of funding and approving studies

Scientists are not free to investigate any scientific question that interests them. They must justify their research and minimise the risk of doing harm. The overall direction of science is controlled by legislation, regulation, policy and funding priorities. There are several regulatory bodies that have oversight of specific areas of scientific research in the UK, e.g. human stem cell research is governed by the Human Fertilisation and Embryology Authority and studies involving patients are approved by the Health Research Authority.

THINGS TO DO AND THINK ABOUT

1. For the following aims, write a possible hypothesis and the relevant null hypothesis.
   a) To investigate the effect of light intensity on the rate of photosynthesis in *Elodea*.
   b) To investigate the effect of listening to different types of music on reaction times of students.
   c) To determine the effect of increasing concentrations of caffeine on heart rate in humans.
2. Use the ethics diagram to create a mind map summarising the various ethical considerations involved in carrying out scientific research.

ONLINE

Find out more about citing and referencing at www. brightredbooks.net

DON'T FORGET

Any definitive statement should be cited and referenced. If you ever find this information is missing, you should question the reliability of the statement.

ONLINE

Find out more about the ethics of human experimentation by reading the Nuremberg Code (1947) at www.brightredbooks.net

SCIENTIFIC LITERATURE AND COMMUNICATION

Science involves the sharing of findings so that data and ideas can be evaluated by others. Scientists do not accept new ideas without being able to evaluate methods, data and its analysis, and conclusions. One of the key concepts of modern science is that sufficient information should be communicated to allow independent **replication** of experiments to **verify results** and **conclusions**.

TYPES OF SCIENTIFIC PUBLICATION

There are two types of formal scientific publication: primary papers and review articles.

Primary papers communicate new research findings and should include:

- an explanatory **title**

- a **summary** containing aims and findings (also known as an **abstract**). The **aim** should clearly state the independent variable (the variable being controlled and deliberately changed), the dependent variable (the variable that the hypothesis states will be affected by altering the independent variable) and the organism, system or molecule being studied

- an **introduction** which explains the purpose and context of the study. Introductions should make use of a variety of sources to support the current study, and must be properly cited and referenced. The introduction should also contain the author's hypothesis

- a **method** section that contains **sufficient information to allow other scientists to repeat the experiments**. The method should clearly define the independent and dependent variables in the study

- a **results** section that includes all relevant **raw data**, **processed data** in the form of summary graphs or tables, and a clear explanation of how the authors have **analysed** the data

- **conclusions** that are supported by the data and refer to the aim and hypothesis. A discussion of the validity and reliability of experimental design should also take place here

- **references** for all cited work

Review articles are used to summarise the current knowledge in a particular field and to discuss recent or novel findings. Review articles are usually written by experts in that area of study and provide an 'up-to-the-minute' account of an area of science.

DON'T FORGET

If a primary scientific publication is lacking any of the described sections, the overall findings of the paper should be treated with caution.

PEER REVIEW AND CRITICAL EVALUATION

A key step in scientific publishing is peer review. This is when scientific papers are read and analysed by specialists with expertise in the field of study. The reviewers assess the scientific quality of the report and the experiments behind it. Often, the paper will not be published without the author first responding to the reviewer's comments. Examples of peer reviewers' requests include:

- alterations to experimental design, such as better control of confounding variables
- increased numbers of replicates

- additional experiments
- different or more suitable data analysis
- inclusion of missing key background information.

Only after successful peer review will papers be accepted and published in scientific journals. However, even if a report has been accepted by reviewers, individual readers should always be **critical** in reading and **evaluating** scientific research.

DON'T FORGET

Scientists should be critical when evaluating any piece of research. If you cannot access or find references, methods or data, you cannot be sure that the conclusions are valid.

SCIENCE REPORTING IN THE MEDIA

Most people do not gain scientific knowledge by accessing formal scientific publications, but instead gain information from the wider media. Scientific breakthroughs that are reported in the media may not have been peer reviewed and, hence, may not have been evaluated by other scientists. In addition, science that is reported in the wider media usually does not include methods and data; sometimes it is not even clear where the report is published or who the authors are. This can lead to misinterpretation of the significance of results and to overhyping of new scientific discoveries.

NEW STERILISATION RULES WOULD COST MILLIONS

Surgery spread AIDS

Could your fridge KILL YOU?

IS SWINE FLU ALREADY HERE?

THINGS TO DO AND THINK ABOUT

1 Access some primary scientific papers using www.ncbi.nlm.nih.gov/pubmed **or** journals.plos.org/plosone/ and attempt to identify the following:
   a) the authors
   b) the authors' institute or university
   c) the time taken to peer review (difference between date submitted and date accepted)
   d) the independent and dependent variables
   e) the aim
   f) the hypothesis
   g) evidence of citations and references in background information.

2 Read a recent newspaper article describing a scientific story. Can you identify from the article the authors of the study, where it was published, the methods used and the data that conclusions are based on?

ONLINE

Find out about the roles played by scientists, the media and society in the recent MMR vaccine scare at www.brightredbooks.net

SCIENTIFIC EXPERIMENTATION: PILOT STUDY AND EXPERIMENTAL VARIABLES

PILOT STUDY

A **pilot study** is a short experiment, or series of experiments, that are commonly carried out before starting a scientific investigation. The purpose of the pilot study is to develop and improve experimental protocols, ensuring that the experimental design is suitable for investigating the aim and hypothesis. The use of pilot studies is necessary to avoid wasting significant time and money on full-scale investigations using flawed experimental methods. A pilot study should be used to provide information and data on a number of parts of experimental design, as summarised on the right.

EXPERIMENTAL VARIABLES

In any biological experiment, there is usually a large number of factors that can change, known as variables. Experimental variables can be split into three types, **independent**, **dependent** and **confounding**.

Independent variable

The **independent variable** is the variable that is deliberately manipulated by the investigator to determine if it has an effect on the outcome of the experiment. Experiments can be **simple**, changing only one independent variable, or **multifactorial**, changing more than one independent variable. In **observational studies**, a variable that already exists within the study population is selected and is considered to be the independent variable (see page 96).

Dependent variable

A **dependent variable** is a variable that is measured to determine if changing the independent variable has an effect.

Ensure validity of experimental design	Validity is the term used to cover the overall experimental concept; to be valid, the study must follow the scientific cycle. Validity includes considering if the experiment has sufficient control groups and appropriate randomisation of experimental subjects. It also includes questioning whether a relationship between the independent and dependent variables supports the hypothesis.
Verification of the effectiveness of techniques	The pilot study allows the investigator to decide if the experimental approach, techniques and equipment are suitable for investigating the aim. The experimental protocol should be evaluated to determine whether measurements are made with an acceptable degree of **accuracy** and **precision**. The pilot study is also an opportunity for the investigator to practise difficult or unfamiliar techniques.
Determine a suitable range of values for the independent variable	It is common to use a linear or log **dilution series** in pilot studies to find out an appropriate range of values for the independent variable. Data from the pilot study allows selection of a range of values for the independent variable that provide an effect on the dependent variable, while avoiding an overly large number of experimental groups.
Identify and control confounding variables	**Confounding variables** are variables, other than the independent variable being tested, that may have an effect on the dependent variable being measured. The pilot study is an opportunity to measure how well these confounding variables are kept constant and to consider if the experimental design has failed to control any important confounding variables.
Identify suitable numbers of replicates	Biological data, by its nature, is variable. In order to be able to come to a **reliable** conclusion, experiments must include a suitable number of **replicates**. These should consist of both replicated measurements within a single experiment and independent replicates from repeated experiments. The pilot study will provide data allowing a decision to be made on the number of replicates required to allow valid analysis of the experimental data.

Importance of a pilot study.

Type of experimental design	Examples	Advantages	Disadvantages
Simple	Commonly laboratory-based studies	● Simpler experimental design ● Variables are tightly controlled ● Results easy to analyse and interpret	● Difficult to extrapolate results to whole organisms or environments ● Not suitable for testing complex hypotheses
Multifactorial	Field studies, whole-organism (*in vivo*) studies	● Allows examination of complex interactions ● Allows generalisations to be made ● Results applicable to study environment or organism	● More complex and time consuming ● Difficult to control all variables ● Analysis of data is complex

Comparison of simple and multifactorial studies.

Confounding variables

A **confounding variable** is a variable, other than the independent variable, that may affect the dependent variable. As biological systems are complex, most biological experiments contain many confounding variables. A valid experimental design should seek to keep all confounding variables constant or, where this is not possible, to monitor confounding variables to allow their effect to be analysed.

If confounding variables cannot be controlled effectively, then a **randomised block design** may be used to minimise the effect of uncontrolled confounding variables.

Example: Randomised block design

In an experiment to test the effect of sex on heart rate in humans, there are many confounding variables that may affect the results, for example age, height, weight, fitness. It would be extremely difficult to control these variables as the researcher would need to find sufficient numbers of subjects all of the same age, height, weight and fitness levels. Therefore, the subjects are grouped (or blocked) randomly, so that the male and female groups contain a similar range of ages, heights, weights and fitness levels. Randomised block design is a form of stratified sampling (see page 99).

Types of variables and data

Variables can be classed as either **discrete** or **continuous**, and give rise to **qualitative**, **quantitative** or **ranked** data.

- **Discrete variables** have finite values that allow placement into separate groups, for example sex, number of pets in a household, or number of students in a class.

- **Continuous variables** change gradually from one extreme to another, potentially having an infinite number of possible values. Examples of continuous variables include height, age, and concentration.

The three different types of data are summarised below.

Table summarising qualitative, quantitative, and ranked data.

Type of data	Explanation	Examples
Qualitative	Descriptive and difficult to measure directly. Recorded using direct counts or by observation. Difficult to analyse. Presented using a bar graph or pie chart.	Colour of leaf, softness of fur, emotions
Quantitative	Measured directly. Recorded as numbers. Easier to analyse. Presented using a line or scatter graph, or a histogram.	Concentration of solution, time, height
Ranked	Data is put into order of magnitude (from smallest to largest or least common to most common). Presented using a bar graph.	Abundance of a plant, dominance hierarchies in animals.

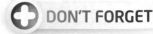

DON'T FORGET

The purpose of a pilot study is to verify experimental protocols and methods, before starting a large-scale investigation. It is normal and expected to use data from a pilot study to alter and improve experimental design.

DON'T FORGET

Biological investigations, especially multifactorial studies, often have more than one dependent variable.

VIDEO LINK

Watch the clip at www.brightredbooks.net to find out more about discrete and continuous variables.

DON'T FORGET

The type of variable being measured determines the appropriate graphical display or statistical test that should be used.

THINGS TO DO AND THINK ABOUT

1   For each of the following experiments list the confounding variables that should be controlled and the type of data that will be obtained.
    a) An investigation into the effect of temperature on the rate of respiration in yeast.
    b) To determine the effect of age on blood pressure in humans.
    c) An investigation of the antimicrobial properties of garlic paste on growth rates of *E. coli*.
    d) The effect of nitrogen concentration on root length of cress plants.

2   Look at some examples of scientific graphs to determine if the data is qualitative, quantitative or ranked, and decide whether the authors have used an appropriate form of presentation.

SCIENTIFIC EXPERIMENTATION: CONTROL GROUPS, OBSERVATIONAL STUDIES AND EVALUATING EXPERIMENTAL DESIGN

ONLINE

Read more about control groups at www.brightredbooks.net

DON'T FORGET

If either the negative or positive control groups give unexpected results, the experiment is not valid and any results obtained cannot be interpreted.

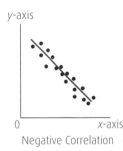

Positive Correlation

Negative Correlation

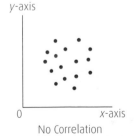

No Correlation

Example of positive and negative correlation. Experimental studies, where the independent variable is directly manipulated by the experimentor, are required in order to identify causation.

CONTROL GROUPS

In any scientific experiment, control groups should be included to enable comparison of the experimental treatment group with the controls. It is only by comparing the treatment group's results with control group results that the experimental design can be shown to be valid. There are two types of experimental control group, **negative controls** and **positive controls**.

Negative control group	The negative control group provides data for what happens in the absence of a treatment. It is carried out to check false positive results are not occurring. The negative control should show that the dependent variable remains stable, within an expected range, when the independent variable is absent or constant. If the negative control gives unexpected results, it is likely that a confounding variable is not adequately controlled.
Positive control group	The positive control group provides data to prove that the experimental design can detect a change in the dependent variable when one occurs. It is carried out to check whether false negative results are occurring. If the positive control shows no effect on the dependent variable, the experimental design is not valid and requires refinement.

Example: Testing the effectiveness of a new disinfectant

In an experiment to test the ability of a new disinfectant to control growth of bacteria, the investigators grow bacteria on an agar plate that is supplemented with the disinfectant. After 24 hours' incubation the number of bacterial colonies is counted. In this experiment, the negative control would be an agar plate inoculated with bacteria but containing no disinfectant. The positive control is an inoculated agar plate containing a different disinfectant that is known to kill bacteria. Expected results are shown below.

Negative control
agar + bacteria +
no disinfectant
Shows normal
bacterial growth.

Treatment group
agar + bacteria +
new disinfectant
Shows some bacterial growth
depending on effectiveness of
new disinfectant and
concentration tested.

Positive control
agar + bacteria +
other disinfectant
Shows little or no
bacterial growth

Example of negative and positive control group expected results.

OBSERVATIONAL STUDIES

In some contexts, such as epidemiology or psychology studies, it can be difficult for investigators to control the independent variable, due to ethical concerns or the impracticality of creating randomised experimental groups. For example, if a scientist wanted to study the effects of smoking on heart disease, it would be unethical, extremely expensive and time consuming to recruit two groups of people and have one group start smoking for a number of years to compare the effects to a control group of non-smokers. Instead, investigators may carry out **observational studies**, using groups that already exist at the expense of being unable to control the independent variable (e.g. smokers and non-smokers).

Observational studies using existing groups are good at detecting **correlations** between two variables. **Correlations** maybe **positive** – as one variable increases the other variable increases too – or **negative**, whereby one variable increases as the other variable decreases.

contd

The dispersion of the points about the line of best fit gives an indication of the variability of the data and the strength of the correlation, if many points are far from the line then the correlation is weak. Although observational studies can show correlations between two variables, care should be taken in assigning **causation**; such studies may not control confounding variables effectively.

Advantages and disadvantages of *in vivo* and *in vitro* studies

In biological studies, experiments generally fall into two types: *in vivo* and *in vitro*. *In vivo* is Latin for 'within the living', and is the term used to describe studies involving living organisms. *In vitro* is Latin for 'in glass', and is the term used to describe laboratory studies using isolated components of organisms, e.g. enzymes or proteins.

Type of study	Advantages	Disadvantages
In vivo	<ul><li>Provide data for effects in whole organisms</li><li>Allows study of complex interactions</li></ul>	<ul><li>Expensive and time consuming</li><li>Ethical and legislative concerns</li><li>Difficult to control confounding variables</li><li>Results may be difficult to interpret</li><li>Difficult to prove causation</li></ul>
In vitro	<ul><li>Simpler and less expensive</li><li>Easier to control confounding variables</li><li>Interpretation of results is simpler</li><li>Can demonstrate correlation and causation</li></ul>	<ul><li>Difficult to extend results to whole organism or different species</li><li>Difficult to model complex interactions</li></ul>

EVALUATING EXPERIMENTAL DESIGN

Evaluating experimental design is a complex activity and requires consideration of a number of factors. As a reader you should assess the experimental design for **validity** and **reliability**. This should include consideration of:

- the suitability of the experimental approach for testing the given aim and hypothesis. If the design is unsuitable, the experiment is invalid

- control groups. **Treatment effects must be compared to effects in control groups**. If either negative or positive controls are missing or give unexpected results, the results must be interpreted with extreme caution and may be dismissed

- adequate **control of confounding variables**. Where confounding variables cannot be controlled, they must be monitored and their contribution to changes in the dependent variable considered

- **selection bias**. This results in non-random and unrepresentative groups and should be avoided. Selection bias happens when the individuals who are selected for a study are not representative of the population as a whole

- **sample size**. This is the number of tests or individuals in treatment and control groups. Sample size should be large enough to allow results to be tested statistically and should aim to be as near to a representative sample as the method allows.

THINGS TO DO AND THINK ABOUT

1 Design suitable negative and positive controls for the laboratory techniques from Unit 1 below:
   a) Colorimetry to determine the concentration of an unknown
   b) Thin-layer chromatography to separate photosynthetic pigments
   c) Separation of proteins by gel electrophoresis
   d) ELISA
   e) Viable cell counting using a vital stain.

2 Write a paragraph to explain why both *in vivo* and *in vitro* studies are used in biology.

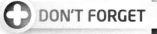

DON'T FORGET

Correlation does not mean that changing one variable *results* in the other variable changing (causation). Only when all confounding variables are controlled, can correlation be proved to demonstrate causation.

ONLINE

Read more about examples of correlations that do not show causation at www. brightredbooks.net

DON'T FORGET

When evaluating experimental design, it is necessary to identify the independent, dependent and all important confounding variables. If the method does not explain how all these variables were measured or controlled, it is likely that the experimental design is not valid.

VARIATION AND SAMPLING

VARIATION

All biological populations show **variation**, that is, differences between individual members of a species. Variation can be either **discrete** or **continuous**. Discrete variation is where there are only a limited number of possible types of characteristic, allowing individuals to be placed into groups – examples include blood group and gender. Continuous variation describes characteristics that change gradually across a range of values, for example height and weight.

Biological variation tends to fall into a **normal distribution**, allowing inferences from a small sample of the population to be extended to the whole population. However, in biology, we see many other types of data distribution, such as skewed.

Normal distribution:

- Also known as a bell curve
- Most common distribution in nature
- Symmetrical
- The central value is the mean, median and mode
- For example, height of humans

Skewed distribution:

- Can be negative (long tail to left) or positive (long tail to right)
- The mean is not the most common value
- The most common value (mode) is not central
- For example, the number of children in a household

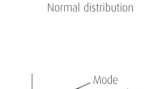

Normal distribution

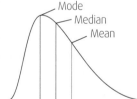

Left-skewed (negative skewness)   Right-skewed (positive skewness)

Skewed distribution

REPRESENTATIVE SAMPLES

In an ideal world, biological studies would gather data from all individuals within a population. However, this is often impractical due to time and cost considerations. Biologists, therefore, conduct experiments on a **representative sample** of a population and seek to extrapolate their results to the population as a whole.

In order for a sample to be representative, it must take into account the extent of natural variation in the population and how the population as a whole is distributed. The greater the variation in the population, the larger the sample size required. Overall, a representative sample should have the **same mean** and **same degree of variation about the mean** as the population as a whole.

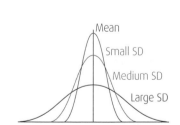

Comparison of low-variation and high-variation population distributions. This figure compares normally distributed data sets with low (blue), medium (black) and high (red) variation. The variation is measured by calculating the standard deviation. A representative sample of the red population would require more individual samples than a representative sample of the blue population.

TYPES OF SAMPLING

There are three main types of sampling strategy: random, systematic and stratified:

- **Random sampling** allows each member of the population to have an equal chance of being selected. It is non-subjective and shows little or no bias in selecting samples. Random sampling should use random number tables or random number generators to select samples. Random sampling is used for sampling very large areas or populations with relatively uniform distributions.

- **Systematic sampling** selects members of a population at regular intervals. For example, in a transect study of plants, a quadrat would be taken at every metre along the transect; when sampling humans, every fifth person in a population could be sampled.

- **Stratified sampling** divides the population into categories that are then sampled proportionally. Stratified sampling is used when the population being sampled is not uniform. The population is made up of subsets of known sizes and the number of samples taken from each subset is proportional to and representative of the whole population. An example of stratified sampling would be measuring the abundance of plants in an area of grassland. Certain plants may grow in patches that might be missed by random sampling, whereas stratified sampling would ensure these areas are sampled proportionally.

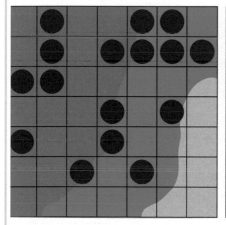
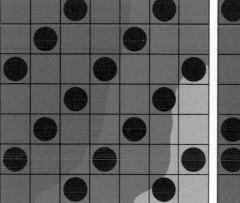

Random, systematic and stratified sampling.

DON'T FORGET

Microsoft Excel has a random number function, simply type =RAN() into a cell.

THINGS TO DO AND THINK ABOUT

1 Learn more about biological data distributions by researching why the distribution of weight in developed countries can no longer be thought of as normal, due to the effects of obesity.

2 Make a table to summarise the advantages and disadvantages of random, systematic and stratified sampling.

ONLINE

Read more about sampling at www.brightredbooks.net

ENSURING RELIABILITY

The natural variability of biological material and populations often leads to variation in data and measurements, especially when studying whole organisms or populations but even with purified molecules. The natural variation in any biological material can be determined by taking measurements from a sample of the population.

Although such variation is to be expected, investigators should also consider the contribution of experimental design and measurement methods in producing variable results. In order to have confidence in results, care should be taken to ensure any measurement methods are suitably **precise, accurate, and reliable**.

DON'T FORGET

To determine the natural variation in a population, a determination of the type of data distribution must be made.

DON'T FORGET

Accuracy and precision are independent; a measuring method can be highly precise without being accurate, and *vice versa*.

PRECISION AND ACCURACY OF MEASUREMENTS

When designing an experiment, consideration should be given to the degree of precision and accuracy required to detect differences in key variables.

- **Precision** is a measurement of the closeness of two or more measurements of the same sample. For example, if you measured the volume of a solution 10 times and get 10 ml each time, then your measurement method is precise. Precision can be assessed by taking multiple measurements and calculating a mean and standard deviation.

- **Accuracy** is an assessment of how close your measurement is to the actual value. For example, if you measure the volume of a 100 ml solution and get a result of 90 ml, the measurement method is not accurate. Accuracy is assessed by calibration against a known standard. So, to assess the accuracy of a balance, the mass of a reference weight of known mass would be measured.

To evaluate precision and accuracy, investigators should assess the following during both the pilot and main studies:

- the mean of replicated measurements
- the variation in these replicated measurements
- the number of samples tested
- how regularly measuring equipment requires calibration against known values.

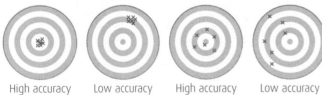

| High accuracy High precision | Low accuracy High precision | High accuracy Low precision | Low accuracy Low precision |

Comparison of accuracy and precision of measurements.

ONLINE

Read more about accuracy and precision at www.brightredbooks.net

RELIABILITY OF RESULTS

Experimental design should use data from the pilot study to determine the number of **replicates** required to ensure a **valid experiment** and produce **reliable results**. Good experimental design should allow for sufficient replicates and independent replicates.

- **Replicates** are multiple measurements within the same experiment, which assess **precision**.

- **Independent replicates** are when the whole experiment is repeated more than once to demonstrate that data is **reliable** and consistent.

DON'T FORGET

The same investigator should carry out the repeated measurements to increase reliability in a valid way.

EVALUATING DATA ANALYSIS

In a results section, data should be presented in a clear and appropriate manner that supports data analysis and interpretation. Replicates in a study provide data that can be used to perform simple statistical calculations, such as the mean, median, mode, standard deviation and range of data, and allow the reader to determine the reliability and validity of a study.

contd

Analysing and presenting variability of data

When presenting data from experiments, it is important to use an appropriate type of graph that also shows the variability of data gathered. The **variability** of data should be shown graphically by plotting **confidence intervals** or **error bars**.

- **Error bars** are lines through a point on a graph, parallel to an axis, which show the variation within the data. Error bars can be used to show direct measures of variation (such as the standard deviation) or probabilities (such as confidence intervals). The smaller the error bars, the less variable the data.

- A **confidence interval** is a statistical estimate of the range of values within which a certain percentage of the total population would be found. For example, a 95% confidence interval shows that the range of values would include 95% of the whole population being studied.

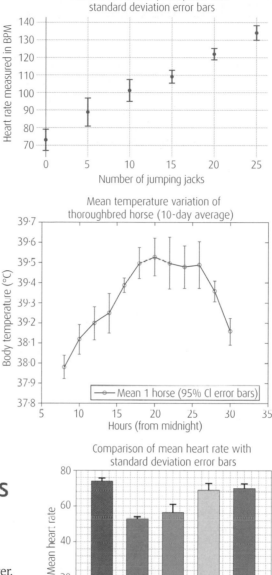

Mean heart rate of participants with standard deviation error bars

Mean temperature variation of thoroughbred horse (10-day average)

— Mean 1 horse (95% CI error bars)

Comparison of mean heart rate with standard deviation error bars

Examples of graphs with error bars displaying variation of data. Can you identify the different ways in which the confidence intervals about the mean have been represented in each case?

EVALUATING DIFFERENCES IN DATA

The aim of scientific studies is to find differences between groups when the independent variable is changed. However, given the natural variability in biological data, it can be difficult to determine if the difference between groups is a valid finding. If the difference between the mean of the treatment group and that of the control group is sufficiently large that their confidence intervals do not overlap, the data can be said to be different.

A **statistically significant result** is one in which the difference between groups is unlikely to be due to chance alone. There are a large number of statistical tests (for example the *t*-test or ANOVA) and the appropriate test to use depends on the type of data and its distribution. The reason for running a statistical analysis is to provide an indication of the likelihood of a difference occurring by chance. Generally, a p **value** $\leqslant$ **0.05** is considered **significant** as there is less than a 5% chance that the difference is due to random variation. Scientists should design experiments that allow significant differences to be demonstrated, by including sufficient replicates and measuring data with appropriate accuracy and reliability.

THINGS TO DO AND THINK ABOUT

Read more about replicates and pseudoreplication at http://bmcneurosci.biomedcentral.com/articles/10.1186/1471-2202-11-5

DON'T FORGET

To avoid pseudoreplication, care should be taken, when designing an experiment that replicates are independent from each other. Pseudoreplication occurs where multiple measurements are taken from the same subject or sample, but are treated as being separate measurements when the data is analysed.

DON'T FORGET

Standard deviation is a statistical calculation used to quantify how much members of a group differ from the mean of the group.

ONLINE

Read more about error bars and statistical significance at www.brightredbooks.net

DON'T FORGET

The type of error bar (usually standard deviation or standard error) that is shown on a graph should be described on the axis label or in the graph legend.

DON'T FORGET

The appropriate form of graph for different types of data is discussed on page 95.

DON'T FORGET

Calculations of statistical tests are not required for the Advanced Higher exam and, although statistical analysis may be helpful for your project, it is not essential.

EVALUATING CONCLUSIONS

At its simplest, a scientific conclusion should state the researcher's interpretation of the meaning of the data gathered in the study. Conclusions should always relate the results directly to the aim and hypothesis of the study. The researcher should use the data gathered, and their analysis of the data, to determine whether:

- changing the independent variable results in changes to the dependent variable
- the findings are statistically significant
- there is sufficient evidence to support or disprove the hypothesis.

When evaluating conclusions, researchers should consider the validity and reliability of the experimental design, whether the results show correlation or causation, and how the findings fit with current scientific understanding and previously published data.

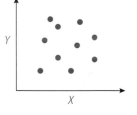

DON'T FORGET

Where scientific studies have been peer reviewed prior to publication, the reader can take confidence that the study *should* be valid and reliable. However, remember that, without independent replication, novel results should be treated with caution.

EVALUATING VALIDITY AND RELIABILITY

Experimental design and presented data should be considered as a whole, to make determinations about the validity and reliability of the study and its conclusions. There should be sufficient replicates and independent replicates to ensure that the data has the required precision and reliability. This can be evaluated by examining the experimental protocol and the variability in results, as shown by error bars. In addition, statistical analysis of the data should support the conclusions. If the data show outliers or anomalous results, these should be discussed and reasons for these results suggested.

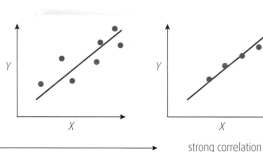

poor correlation   strong correlation

EVALUATING CORRELATION AND CAUSATION

Evaluating the strength of **correlations** between two variables can be achieved by examining the data and by considering how closely the two variables respond to each other. Strong correlations often show a linear relationship, the strength of which can be shown by carrying out a regression analysis. The strength of the correlation is judged by the closeness to the regression line of the individual data points.

Demonstrating **causation** is more difficult as correlation does not prove that the variables show a cause-and-effect relationship. Causation can only be shown where all confounding variables are adequately controlled and so is difficult to demonstrate in observational studies. A well-designed experimental study may allow conclusions to be drawn about causation.

ONLINE

Read more about causation and correlation at www. brightredbooks.net

DON'T FORGET

Scientists should be aware of bias and should not give undue prominence to one set of results over another, just because they support the hypothesis.

CONSIDERATION OF CURRENT KNOWLEDGE

When reaching conclusions, scientists should consider previous studies that have investigated similar or related phenomena. Results should be discussed in the context of previous studies, including references to previously published studies. Where new results conflict with previous studies, reasons for such differences should be explored with reference to the methodology used. Negative results from the current or previous studies should also be considered and discussed. In the case of conflicting results, it may be appropriate to suggest further experiments to test a new hypothesis that would resolve these differences.

THINGS TO DO AND THINK ABOUT

Using your knowledge of investigative biology, read the following study and answer the questions.

(Adapted from: Lalor M.K., Floyd S., Gorak-Stolinska P., Weir R.E., Blitz R., Branson K., *et al.* (2011) BCG Vaccination: A Role for Vitamin D? PLoS ONE 6(1): e16709. doi:10.1371/journal.pone.0016709)

BCG VACCINATION: A ROLE FOR VITAMIN D?

Introduction

BCG vaccination is administered in infancy in most countries with the aim of providing protection against mycobacterial infections such as tuberculosis and leprosy. Vitamin D deficiency is now recognised as widespread [Matsuoka *et al.* 1995] and is more common in TB patients than controls [Wejse *et al.* 2007]. Patients with tuberculosis have, on average, lower blood concentrations of vitamin D than healthy controls [Nnooaham and Clarke, 2008].

In the last five years there has been renewed interest in the biological effects of Vitamin D on tuberculosis due to the growing evidence of the immunomodulatory properties of Vitamin D. Vitamin D alone has no direct anti-mycobacterial action, but its active metabolite 1α25(OII)2D improves the host response to mycobacterium infection [Martineau *et al.* 2007]. Vitamin D has also been shown to have effects on immune cells *in vitro*. Paradoxical effects of vitamin D have been observed in immunity of tuberculosis: vitamin D decreased immunity, but increased bactericidal activity [Cantorna *et al.* 2008].

In this study, we measured Vitamin D, in blood samples from UK infants who had received BCG vaccination and in age-matched infants who had not been vaccinated. We aimed to determine if there was an association between circulating vitamin D concentrations and BCG vaccination status (determined by measuring levels of the immune cytokine IFNγ) in UK infants.

Methods

Blood samples were collected from UK infants who were vaccinated with BCG at 3 ($n = 47$) and 12 ($n = 37$) months post BCG vaccination. These two time-points are denoted as time-point 1 and time-point 2. Two blood samples were also collected from age-matched unvaccinated infants ($n = 32$ and 28 respectively), as a control group. Blood vitamin D concentrations were measured by radio-immunoassay. The cytokine IFNγ was measured in supernatants from diluted whole blood stimulated with *M. tuberculosis* (M.tb) PPD for 6 days.

Results

The median concentration of Vitamin D in all the UK infants measured at time-point 1 was 28.7 ng/ml, which decreased significantly by time-point 2 when the median was 18.9 ng/ml ($p<0.0001$, Chi-squared test). 58% of infants had some level of vitamin D deficiency (<30 ng/ml) at time-point 1, and this increased to 97% 9 months later (Figure 1). BCG-vaccinated infants were almost 6 times (Confidence Interval: 1.8–18.6) more likely to have sufficient vitamin D concentrations than unvaccinated infants at time-point 1 ($p = 0.01$, Chi-squared test) (Figure 1), and the association remained strong after controlling for season of blood collection, ethnic group and sex. Among vaccinees, there was also an association between IFNγ response to M.tb PPD and vitamin D concentration, with infants with higher vitamin D concentrations having lower IFNγ responses (Figure 2).

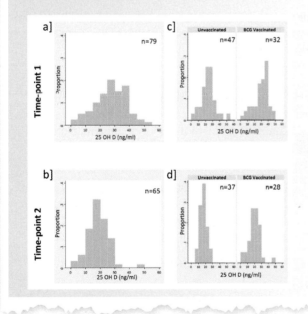

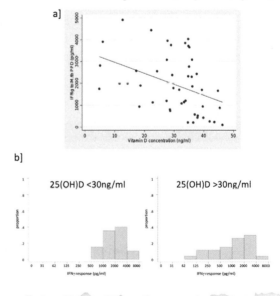

a) What evidence do the authors supply to support the statement that vitamin D deficiency is widespread?

b) What is the aim of this study?

c) Suggest a suitable hypothesis for this study.

d) What type of study is this?

e) What was the control group?

f) Which confounding variables did the study consider when comparing vitamin D levels in vaccinated and unvaccinated children?

g) How have the authors supported their assertion that vaccinated infants were more likely to have sufficient vitamin D levels?

h) Describe the association between IFNγ response and vitamin D concentration.

i) With reference to the methods and results, discuss if sample numbers are sufficient to make conclusions about the relationship between immunisation and vitamin D levels?

j) The authors suggest carrying out a further randomised controlled study to confirm the effects observed. Briefly describe how this study might be carried out.

YOUR INVESTIGATION

CHOOSING A TOPIC

The Investigation provides opportunities for you to put into practice your investigative skills, honed during the taught unit. It must be a piece of **individual work**, so group work and joint investigations are not permitted.

Your teacher/lecturer will help you to choose a suitable topic that interests you. It could be something that you have covered in the AH Biology course, though it doesn't have to be; there may be something that you learnt in Higher or found out about from another source that leads you to a suitable topic for an investigation. Useful sources for initial reading include school and undergraduate textbooks, newspaper and media items, medical and government reports, websites and scientific journals such as *School Science Review*, *New Scientist* and *Scientific American*.

A suitable topic

You may have a great idea for an investigation topic, but it needs to be suitable to your situation. What does this mean?

- Obviously your investigation has to be on a **biological topic** and it must be at a level appropriate to the **demands** of Advanced Higher Biology.

- It must include **experimental or survey work** and can't be purely a technical exercise in statistics or computing. (And if you are doing AH Chemistry or AH Geography, for example, the topics of your investigations need to be different.)

- Do not be over-ambitious. Your investigation is unlikely to be a piece of original research, but it will be **new to you**. A **simple investigation**, able to be completed in the time available, is likely to be the most successful.

A practicable topic

During discussions with your teacher/lecturer, consider the **time and the resources** available to you to gather sufficient data (including the replicates or sample size).

Biological context	Independent variable	Dependent variable
What is it that you plan to investigate?	What will you change in the biological system?	What will you look for as your result?
For example, an enzyme, tissue, organism, habitat, etc.	How will you change the conditions?	How will you measure the result?

Limitations might include: equipment, biological materials, chemicals, safety, laboratory facilities, season of year.

You need to think about the **aim of the Investigation** and how you will gather results. The aim has three parts and they all need to be practicable for your investigation to be feasible.

Making a timeline

When you have sorted out what experimental or survey work you will do to investigate your topic, you need to make a realistic timeline of the work that you plan. You will need a good level of **self-motivation and organisation** so that you can stick to the timeline. You may have to be flexible – your plans may have to change as your work progresses. Making an early and committed start to the work will allow time for this flexibility; if you start late, the final report deadline may catch up with you.

KEEPING A RECORD

Keeping a regular, clear and accurate record of your investigation work is crucial to producing a good report. You should also record your results and other data in a way that makes it easy for you to follow and understand. It is much easier later if you note the **date of each entry** in your record!

contd

Your record should include:

- short notes on the ideas you have in the early stages of planning your investigation (including ideas you might later dismiss)
- a brief record of the discussions you have with your teacher/lecturer and other scientists (including any decisions that you make)
- notes from your background biology research into the topic of the Investigation (including reference details)
- a record of experimental methods and procedures, results obtained and an analysis of your results (using calculations, tables, graphs, diagrams or photographs as appropriate). This would include any preliminary work carried out as a pilot study (e.g. developing techniques or apparatus; devising observational, experimental or sampling procedures).

Periodically, your teacher/lecturer will check on how you are progressing in your investigation and they will want to see your record. These checks will happen regularly to ensure that procedures, safety, resources and timescales are still appropriate. Although you are expected to work autonomously, your teacher/lecturer will provide advice and support and, if these discussions lead to a change in your plan, it is important that you keep a note of the reasons for any changes.

THE PILOT STUDY

To pass the Investigative Biology unit in full, you are required to plan and carry out a pilot study for your Investigation, as well as to pass the written test. This pilot study could be assessed from the notes in your investigation record or as a separate written report. (If you have already passed the knowledge assessments for the other two units, passing the pilot study will also allow you to pass these units in full.)

The table below shows what you need to do to pass the pilot study.

Checkpoint	Suggestions on how to achieve the checkpoint
1. State the aim of the Investigation and formulate questions or hypotheses to be investigated	Develop ideas for your investigation by reviewing and discussing previous learning and/or researching appropriate sources of information. The purpose of your investigation must be clear and you must formulate questions to be investigated and/or hypotheses to be tested.
2. Devise appropriate experimental, observational and sampling procedures, techniques and apparatus	The procedures you devise must be appropriate to the aim of the Investigation. You must select an appropriate procedure after considering or trying alternatives or becoming proficient in the procedure.
3. Consider the need for controls and replicate treatments or samples	You must consider the use of negative and positive controls and the control of potential confounding variables as appropriate. The need for repeat measurements, replicate treatments or samples and independently replicated experiments must be considered.
4. Take into account the ethical use of living materials, human subjects and the conservation of natural habitats	Take into account any ethical issues relevant to the Investigation.
5. Identify potential hazards, assess their associated risks and apply appropriate control measures	Be aware of any potential hazards and use the appropriate control measures to control risks in carrying out the Investigation.
6. Make observations and record measurements with appropriate precision and accuracy	Record your observations and/or measurements in a planned and organised way. Consider the precision and accuracy of your results.
7. Use initial results to devise further experiments or to confirm the appropriateness of a procedure for further work	Review experimental findings and identify if further steps are needed.

THE WRITTEN REPORT

The final report is sent to the SQA. It has to be written up in a formal style and is worth 30 of your 120 total marks. The latest guidance on how to present the report (along with the detailed marking scheme) is found at **Coursework Information** on the SQA AH Biology website: http://www.sqa.org.uk/sqa/48458.html

WRITING THE REPORT

MAKE SURE YOU FOLLOW THE INSTRUCTIONS

The Investigation Report submitted to SQA is worth 25% of the final Advanced Higher mark. It is worth ensuring that your report matches the examination requirements as closely as possible; double check using the web reference on the previous page.

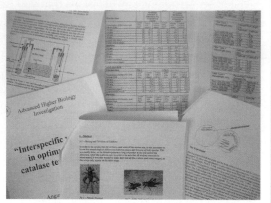

An investigation report should be structured logically and written clearly. The main body of the text should be around 2000 to 3000 words.

Remember that an examiner will have to mark your work, so make it as **logical** and as easy to read as possible. Communicate **clearly** and honestly, and with enough **detail** so that your investigation could be repeated by anyone who has read your report.

Make sure that all your **raw data** is included; if there is a lot of data, include it as an appendix. The rest of this section offers advice in areas where candidates frequently and unnecessarily lose marks.

The report must include the following sections:

- Title page
- Contents page
- Abstract
- Introduction
- Procedures
- Results
- Discussion
- Reference list.

The detailed marking scheme for the Investigation Report and guidance on what to write for each section of the Investigation report can be is found at **Coursework Information** on the SQA AH Biology website; the link can be found at www.brightredbooks.net

YOUR INTRODUCTION

Your introduction must include a clear statement of the aims of the Investigation (despite the fact that you have already stated these in the abstract), along with relevant hypotheses or questions.

In this section, you must also include an account of the relevant background theory at a level appropriate to Advanced Higher. You are expected to provide plenty of detail. Remember, the examiner may know little about your chosen topic, so your introduction should provide the background information. You must also justify the biological importance of the Investigation. Why did you think it was worth doing in terms of biological knowledge? Does the topic have wider implications?

GRAPHICAL PRESENTATION

It is essential to summarise your results using appropriate graphs. But graphs should not appear alone; every graph that you draw should be accompanied by a table showing the relevant processed data.

You may be adept at using electronic software to manipulate the axes of graphs to suit your data, so that the results are presented in a correct scientific manner. If you are at all unsure, you are advised to draw your graphs by hand. Check your graphs to make sure they present your data clearly:

contd

- Are the axes labelled, including units?
- Is there a zero at the start of each axis?
- Are the scales evenly spaced?
- Have you used line graphs for continuous variables and bar graphs/pie charts for discontinuous variables?
- Have you presented your data sets so that they can be compared easily?

YOUR CONCLUSIONS AND EVALUATION

The discussion section is the most important part of the Investigation Report. You must refer back to what you have written earlier in the report and discuss your findings in a critical and scientific manner.

In your discussion section, you need to provide a clear statement of the **overall conclusion**, stating how your results relate to the biological effect you were investigating. Your conclusion must relate to the aim of the Investigation and be valid for the results obtained. Do not be tempted to over-interpret what you have found!

Essentially, this section is your chance to explain whether your independent variable has influenced your dependent variable. In order to do this well, you should evaluate whether your **procedures** were robust enough for you to be able to detect an effect – or are there issues to do with the accuracy, replication, sample size, controls or errors? You must also explain very carefully how confident you can be about your results. Your discussion should cover the degree of **variation** in your results **within samples**, **between treatments** and, finally, **between replicates**.

REFERENCES

A **reference** is any piece of material to which a writer '**refers**' in the text. For example, the authors of this book (Lloyd and Morgan, 2016) find that they often refer to Raven and Johnson (2002) and Begon *et al.* (2006) when planning their teaching. The journal *Biological Sciences Review* is also recommended as it contains many articles

Scientists frequently refer to publications during investigative work.

relevant to Advanced Higher Biology, such as one on growing algae for food (Roberts, 2007). Website references must include the following where possible: author, date of publication, title, publisher, URL and the **date you accessed the material** (because the website may be updated later).

Example: Reference list

Begon, M., Townsend, C.R. & Harper, J.L. (2006) *Ecology: From Individuals to Ecosystems (4th Edition)*, Blackwell Publishing, Oxford
Lloyd, D. & Morgan, G. (2016) *Advanced Higher Biology,* Bright Red Publishing, Edinburgh
Raven, P.H. & Johnson, G.B. (2002) *Biology (6th Edition)*, McGraw-Hill, New York
Roberts, D. (2007) Aquaculture *Biological Science Review* Vol. **20** (No. 1), pages 2–6
Scottish Qualifications Authority (2015) *Biology Project report General Assessment (valid from session 2015–2016 and until further notice)*, Scottish Qualifications Authority URL: http://www.sqa.org.uk/files_ccc/GAInfoAHBiology.pdf
(date accessed October 2016)

DON'T FORGET

You could use this style of referencing for Advanced Higher Biology.

APENDICES

MODEL ANSWERS

SCIENTIFIC METHOD AND ETHICS (PP. 90-1)

1. a) Hypothesis example: As light intensity increases rate of photosynthesis in elodea will also increase.

Null hypothesis: Changing light intensity has no effect on rate of photosynthesis.

b) Hypothesis example 1) Types of music with high tempos will increase reaction time more than slow tempo music.

Hypothesis example 2) Music with lyrics will increase reaction times more than instrumental music.

Null hypothesis: The type of music being listened to will have no effect on reaction time.

c) Hypothesis example 1): Increasing the concentration of caffeine consumed will result in increased heart rate in humans.

Hypothesis example 2) Increasing the concentration of caffeine consumed will result in decreased heart rate in humans.

Null hypothesis: Changing the concentration of caffeine consumed will have no effect on heart rate.

SCIENTIFIC LITERATURE AND COMMUNICATION (PP. 92-3)

1. Model answer for paper found at http://journals.plos.org/plosone/article?id=info%3Adoi%2F10.1371%2Fjournal.pone.0156112

a) Authors: Jean-Nicholas Audet, Simon Ducatez and Louis Lefebvre.

b) Affiliation: Department of Biology, McGill University, Montréal, Québec, Canada

c) Date Submitted: April 14 2016, Date accepted June 23, 2016, Peer review time 2 months

d) Independent variable: ability to pass 'string-pulling test'; Dependent variables: performance on other behavioural

measurements (shyness, neophobia, problem-solving, discrimination, reversal learning performance).

e) Aim: To investigate how sting-pulling performance correlates with performance in other behavioural tests in Barbados bullfinches and Carib grackles.

f) Hypothesis: Not explicitly stated. Possible hypothesis is that string-pulling results will correlate with problem-solving scores but not discrimination and reversal learning tests. (see Introduction paragraph 4, final sentence)

g) Citations throughout text in Vancouver (numbered) style. Full reference list provided at end of article.

SCIENTIFIC EXPERIMENTATION: PILOT STUDY AND EXPERIMENTAL VARIABLES (PP. 94-5)

1. a) Confounding variables: Type of yeast, duration of experiment, method of measuring respiration, concentration of yeast in suspension, diluent used to make suspension, pH of yeast suspension, concentration of respiratory substrate (sugar) provided, type of respiratory substrate provided.

Type of data: quantitative

b) Confounding variables: Sex, height, weight, fitness/weekly exercise levels, smoking status, time of day, food and drink consumed prior to experiment, blood pressure monitor used, person measuring blood pressure, temperature of room where tested, position when tested (sitting/standing etc), time resting prior to test, duration of test, distractions during test (quiet or loud environment, subject talking or silent etc.).

Type of data: quantitative

c) Confounding variables: Strain of E. coli used, incubation temperature, incubation time, type/source of garlic paste, growth medium used, methods of measuring bacterial growth.

Type of data: quantitative if number of bacterial cells or cell concentration is measures; qualitative if degree of coverage of plate is assessed.

d) Confounding variables: source of nitrogen (e.g. type of fertiliser or chemical), growth medium, volume of water, source/type of cress seeds, number of cress seeds, duration of experiment, temperature of incubation, pH of growth conditions, light intensity, person and method of measuring root length.

Type of data: quantitative

SCIENTIFIC EXPERIMENTATION: CONTROL GROUPS, OBSERVATIONAL STUDIES AND EVALUATING EXPERIMENTAL DESIGN (PP. 96-7)

1. a) Colorimetry to determine the concentration of an unknown molecule. Negative control would be tube with dye but none of the unknown molecule. Positive control would be tube with dye and concentrated sample of unknown molecule.

b) TLC to separate photosynthetic pigments. Negative control would be a sample containing only buffer used to purify pigments from leaf. Positive control would be a sample known to contain expected photosynthetic pigments.

c) Separation of proteins by gel electrophoresis. Negative control would be a sample containing buffer but no protein. Positive control would be a sample containing proteins of known molecular weights (e.g. molecular weight markers).

d) ELISA. Negative control would be wells containing conjugated enzymes and detection substrate but no protein. Positive control is wells containing antibodies, a known concentration of purified protein being tested and detection substrate.

e) Viable cell counting using vital stain. Negative control would be a sample of cells with no vital stain. Positive control would be a sample containing a mixture of live and dead cells stained with vital stain.

2. In vivo studies:
- allow collection of data of effects on whole organisms
- allow for examination of complex interactions, e.g. how treatment effects multiple cell types and their interactions.
- In vitro studies:
- simpler to design experiments and control confounding variables
- less expensive
- less time consuming
- more straight forward to interpret results and show causation
- fewer ethical and legislative concerns

VARIATION AND SAMPLING (PP. 98-9)

1. Historically weight distribution has been seen to be normal with the excepted bell shaped curved about the mean. However, the easy and cheap availability of high calorie food and the reduced activity levels found in developed countries has led to a right-skewed distribution as well as an increased mean BMI. For more information see http://www.nature.com/ijo/journal/v25/n10/full/0801715a.html

2.

Type of sampling	Advantages	Disadvantages
Random	• Non-subjective • Little or no selection bias • Useful for sampling large areas or populations • Straight forward to select individuals/areas to be sampled using random number tables/generators	• Requires population to have a uniform distribution • May lead to poor representation of population or large areas being underrepresented depending on random numbers generated
Systematic	• Simpler methodology than random sampling • Sampling is a uniform intervals • Ensured sample area/population is evenly covered	• Increases chance of bias occurring as not all area/individuals have even chance of being sampled • Can result in over or under representation of areas/individuals
Stratified	• Can be combined with random or systematic approaches • Allows sampling of non-uniform populations • Can generate more representative results • More flexible approach	• Relative distributions must be known and accurate

EVALUATING CONCLUSIONS (PP. 102-3)

a) The authors cite an article by Matsuoka et al. 1995 to support the statement that Vitamin D deficiency is widespread.

b) The aim of the study is to determine if there is an association between circulating vitamin D concentrations and BCG vaccination status.

c) Although the authors do not state a hypothesis the introduction suggests that a likely hypothesis is that BCG vaccination status will be positively correlated with Vitamin D levels. The null hypothesis would be that BCG vaccination status has no effect on vitamin D levels.

d) This is an ex vivo, observational study using a stratified sampling approach.

e) The control group was age-matched unvaccinated infants.

f) The study controlled for age, season of blood collection, ethnic group and sex.

g) To support the conclusion that vaccinated infants were more likely to have sufficient vitamin D levels the authors provide Confidence Intervals and a significant (p=0.01) statistical analysis (Chi-squared test).

h) INF-γ and vitamin D concentration show a negative correlation, as Vitamin D concentration increases IFN-γ tends to decrease. The correlation relatively weak with many points falling relatively far from the correlation line.

i) The study used relatively few infants in each group (47 and 37 for vaccinated infants and 32 and 28 for control unvaccinated infants). Study numbers were sufficient to show a significant link between sufficient vitamin D levels and vaccination status. There is no testing of the strength of the correlation between vitamin D levels and IFN-γ response and the correlation looks weak according to Figure 2. Therefore for the first conclusion numbers are sufficient but for the second conclusion they are likely to be insufficient.

j) The authors would conduct a randomised controlled study by taking infants and randomly assigning them to two groups. One group (treatment) would be vaccinated with BCG vaccine and the other group (control) would receive either not vaccine or a placebo injection containing no vaccine. The authors would then measure vitamin D levels to determine if vaccination has had an effect.